The Magical Maze

The Magical Maze

SEEING THE WORLD
THROUGH MATHEMATICAL EYES

Ian Stewart

Weidenfeld & Nicolson
LONDON

First published in Great Britain in 1997
by Weidenfeld & Nicolson

© 1997 Ian Stewart

The moral right of Ian Stewart to be identified as the
author of this work has been asserted in accordance with
the Copyright, Designs and Patents Act of 1988

A CIP catalogue record for this book is available
from the British Library.

Typeset by
Selwood Systems
Printed in Great Britain by
Butler & Tanner Ltd. Frome and London

Weidenfeld & Nicolson

The Orion Publishing Group Ltd
Orion House
5 Upper Saint Martin's Lane
London, WC2H 9EA

CONTENTS

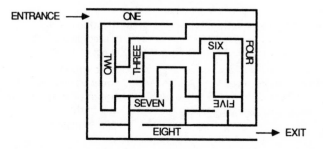

ILLUSTRATIONS

Figure 77 H.-O. Peitgen, H. Jürgens, and D. Saupe, *Chaos and Fractals*, Springer-Verlag, New York, Plate 3, fol. p. 152.

Figure 83 Przemyslaw Prusinkiewicz and Aristid Lindenmayer, *The Algorithmic Beauty of Plants*, Springer-Verlag, New York, 1990, Fig. 1.24, p. 25.

Figure 84 H.-O. Peitgen, H. Jürgens, and D. Saupe, *Chaos and Fractals*, Springer-Verlag, New York, Fig. 5.48, p. 280.

Figure 85 Michael F. Barnsley and Lyman P. Hurd, *Fractal Image Compression*, A.K. Peters, Wellesley, MA, 1993, Plates 13 and 14, p. 53.

Figure 86 Ian Stewart, *Game, Set & Math*, Blackwell, Oxford, Fig. 9.2, p. 127.

Figure 87 Edward Ott, *Chaos in Dynamical Systems*, Cambridge University Press, Cambridge, Fig. 1.13, p. 15.

BEFORE YOU ENTER ...

S ome scientists talk to the people; most, deplorably, don't. One who did was Michael Faraday, one of the greatest scientists of the nineteenth century. Faraday made enormous advances in the theories of electricity and magnetism – in particular, he invented the electric motor and the dynamo. He provided the foundations upon which James Clerk Maxwell built his masterpiece, the mathematical equations of electromagnetic fields. From Maxwell's work it was but a short step to the discovery of electromagnetic waves, from which – thanks to numerous mathematicians, physicists, engineers, inventors, and entrepreneurs – came radio, radar, and television.

The television connection closes a curious historical loop. Faraday's career is intimately bound up with the Royal Institution of Great Britain, a building in London that housed its own scientific library, laboratories, and lecture theatre. He first became interested in electricity while he was working as an apprentice bookbinder, and read an article on the topic in the third edition of the *Encyclopaedia Britannica*. He was given a ticket to attend a lecture at the Royal Institution, given by the great chemist Sir Humphry Davy. The young Faraday was spellbound. He wrote to Davy asking for a job, and when one of the great man's assistants was sacked for getting into a fight, Faraday became Davy's laboratory assistant. By 1820 he knew as much chemistry as anyone alive – but his attention was turning to electricity once more.

In 1821 Faraday married Sarah Barnard, and settled permanently at the Royal Institution. For the next thirty years he carried out his epic work on electricity and magnetism, along with much else. And he did not neglect the general public. In 1826 the Royal Institution began a tradition that continues to this day: the Christmas Lectures for young people. They have been held every year since, with one short break during the Second World War. Faraday gave nineteen of the lectures between 1827 and 1860.

And here is where the circle closes, for in recent years the British Broadcasting Corporation has televised the Christmas Lectures.

The Magical Maze came into being because Professor Peter Day, Director of the Royal Institution, invited me to give the 1997 Christmas Lectures – the 168th lectures in the series, and only the second time (shame!) that they have focused on mathematics. The opportunity to produce a book, based around the lectures and published simultaneously, proved irresistible. Of course the book had to be written before the lectures were given; moreover, the style of a book differs somewhat from the style of a lecture. So *The Magical Maze* treats the material in a rather different manner, with descriptions and pictures replacing demonstrations, apparatus, and interactive sessions with members of the audience. It is aimed at anyone who is interested in mathematics, not just 'young people'. The lectures themselves are drawn from roughly half of the book: the other half is a bonus. (We haven't, at the time of writing, decided exactly which topics will appear in the lectures, so I can't tell you *which* half. Sorry.)

When an author first conceives of a book, it is a ghostly, shimmering wraith – a virtual object. It doesn't yet exist. So many things *could* go in, so many words *could* be placed on the page … As the writing progresses, the ghost becomes more solid, the virtual becomes more real. Words, paragraphs, chapters come into being.

There are many choices, many decisions. What to include? What to exclude? Each decision carries consequences: early material cannot be omitted if later material depends on it. The possible structures for the book are an interconnected network.

And yet every book must tell a story. A story has a beginning, a middle, and an end. It is read as a linear sequence.

When a book is being written, it is a maze of possibilities, most of which are never realised. Reading the resulting book, once all decisions have been taken, is like tracing one particular path through that maze. The writer's job is to choose that path, define it clearly, and make it as smooth as possible for those who follow.

Mathematics is much the same. Mathematical ideas form a network. The interconnections between ideas are logical deductions. If we assume *this*, then *that* must follow – a logical path from *this* to *that*. When mathematicians try to understand a problem, they have to thread a maze of logic. The body of knowledge that we call mathematics is a catalogue of interesting excursions through the logical maze.

This is why I have chosen to use the metaphor of a maze to describe mathematics. As for its magical nature … well, we'll get round to that shortly. It is also why I have structured *The Magical Maze* as one particular journey through the verbal maze of books that might have been instead. Instead of a preface or prologue, I've started with an Entrance. Instead of reading chapters, we wander through Passages. Passages meet at Junctions. In place of a conclusion, climax, or epilogue, we finish at the Exit.

And on at least one occasion, we run smack bang into a Dead-end.

You don't need to thread the maze yourself: my job is to guide you through that one selected path that crystallised out when vague thoughts transmuted into words on paper.

I must admit that I nearly didn't have a 'Before You Enter' section – my term for a foreword – at all. However, I needed to explain about mazes *before* I propelled you into one. And I also needed to do various things that most authors do in forewords, such as telling you why the book ever got written, which I've now done, and thanking the people who helped make the book possible, which I'm just getting round to. Foremost among those people is Peter Day, without whose interest I would never have had the opportunity to put together five television programmes on mathematics. Ravi Mirchandani at Orion Publishing deserves considerable credit for his enthusiasm and guidance, as does Benjamin Buchan. My debt to Caroline Van den Brul and Martin Mortimore of the BBC is so vast that I scarcely dare acknowledge it in print. Professor Sir Brian Follett, Vice Chancellor of the University of Warwick, graciously redefined my duties to make time for me to engage with the public. Alan Newell, Chairman of the Mathematics Department, was an enthusiastic supporter of this development, which now takes tangible form as MAC@W – the Mathematics Awareness Centre at Warwick.

Finally, I owe a long-standing debt to Professor Sir Christopher Zeeman, founding father of Warwick's Mathematics Institute and for many years my boss. Christopher was the first mathematician to give the Christmas Lectures; he also influenced my career in many ways, in particular by encouraging me to indulge in non-academic activities aimed at bringing new mathematics to the people.

And we are all indebted to Michael Faraday, without whom neither the Christmas lectures, nor television, would have been invented.

I.N.S.

Singapore, Australia, New Zealand, Hawaii, Sweden, and Coventry
January–May 1997

ENTRANCE

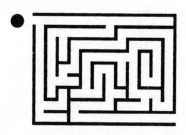

Welcome to the maze.

A logical maze, a magical maze. A maze of the mind.

The maze is mathematics. The mind is yours. Let's see what happens when we put them together.

What is mathematics? What mathematicians do.

What is a mathematician – someone who does mathematics?

Not exactly. That's too easy an answer, and it creates too simple a maze – a circular loop of self-referential logic. No, a mathematician is more than just somebody who *does* mathematics. Think of it this way: what is a businessman? Someone who does business? Yes, but not just that. A businessman is someone who sees an *opportunity* for doing business where the rest of us see nothing: while we're complaining that there's no restaurant in the area, he's organising a telephone pizza delivery service. Similarly, a mathematician is someone who sees opportunities for doing mathematics that the rest of us miss.

I want to open your mind to some of those opportunities.

I don't want to convince you that mathematics is useful. It is, but utility is not the only criterion for value to humanity. I *do* want to convince you that mathematics provides a lot of fascinating insights into the natural world – including bits of it that most of us seldom connect with mathematics, such as the shapes of plants, the markings on animals' coats, and how living creatures move around. Above all, I want to convince you that mathematics is beautiful, surprising, enjoyable, and interesting. You can get a lot out of mathematics for its own sake, and historically, that's where a lot of it came from. The rest came from a two-way trade between the natural world and the human mind.

Some famous mathematicians (no names, no pack drill) have argued that the only good mathematics is what they call 'pure' mathematics, which they praise for its lack of utility. I think that's nonsense, as well as being an intellectual pose of the worst kind. Being useless is nothing to be proud of. But to me, nuts-and-bolts applications to human affairs are the icing on the mathematical cake: they enhance the interest, the surprise, the fun, and the beauty – but they're not the *reason*. As far as *The Magical Maze* is concerned, the idea is to appreciate the cake *as* cake, whether or not there's any icing – and whether or not you get a bit of icing on your plate.

The title *The Magical Maze* sums up many of my feelings about that kind of approach to mathematics. It is a maze because being a mathematician involves navigating, with confidence, through an intricate network of logical possibilities. At every step in a mathematical investigation you are faced with choices. Some choices lead in fruitful directions; most do not. So mathematics is a maze: a logical one.

It is also magical. In fact, mathematics is the closest that we humans get to true magic. How else to describe patterns in our heads that – by some mysterious agency – capture patterns of the universe around us? Mathematics connects ideas that otherwise seem totally unrelated, revealing deep similarities that subsequently show up in nature. Even though mathematics is 'just' a creation of the human mind, it has given us enormous power over the world we inhabit. In mathematics, you can set out to understand the notes played by a violin, and end up inventing television.

That's *real* magic.

In *The Magical Maze* you will encounter many different kinds of mathematics, in many different circumstances. A lot of those circumstances are puzzles and games – a tradition that goes back to the dawn of our subject. In ancient Babylon, some four thousand years ago, the mathematical priests and scribes taught their subject through puzzles. But despite their surface appearance, these games and puzzles are far from trivial. Part of the magic of mathematics is how a simple, amusing question can lead to deep and far-ranging insights.

We'll start with some simple magic, the kind of trickery that goes down well at parties. We'll take a look at mazes, too. But by the time we've finished, you will have learned to thread the *real* magical maze, the mind-world of mathematics, an unseen world of boundless opportunity. You will have discovered what it is like to think like a mathematician. And you will see the world around you with new eyes, eyes that are open to the beauty of mathematics and its relation to the beauty of nature.

At least, that's my intention.

And unless you make the journey, you'll never find out what's at its end.

The magical maze awaits.

JUNCTION ONE

*F*rom *a distance, the magical maze looks drab and uninviting. Most of it seems to be built from faceless concrete slabs, except for some kind of tower that rises above it, almost hidden by the perpetual mists that engulf the structure. Squads of faceless subhumans labour in its shadow, making piles of stones and recording their numbers on slates.*

The mists clear for an instant, and the tower's surface flashes in a shaft of sunlight. It is made from something shiny.

Ivory?

Before you can decide, the mists close again.

As you approach the colossal structure, however, it slowly seems to change. It is as if its magic is so great that some of it is leaking out. No longer a concrete monolith, it begins to reveal artistic embellishments – curlicues and trellises, mythological gargoyles, huge buttresses of fossil-bearing stone. The faceless subhumans flee into concealed burrows. Their places are taken by resplendent beings who radiate serenity and wisdom.

The tower peeps through again, but now it seems to be made not of ivory, but of brushed aluminium.

Closer still, and the wealth of detail begins to be overwhelming. The magical maze possesses an almost tangible aura. Its quality is surreal. The more ordinary it looks, the more remarkable it feels.

You could swear it was growing as you watched.

The path passes through a series of gates, like canal locks. Signposts decorated with esoteric symbols point in every direction. You are about to panic when you see one that says ENTRANCE THIS WAY.

You climb a long spiral staircase to a broad piazza, surrounded on three sides by vast stone walls. Strange cabbalistic designs cover its surface – they draw you in, compulsively, towards a place where the stones have collapsed to leave a small opening.

You clamber through. You are in some kind of garden. At the same time, you are in no doubt that it is a maze. The magical maze.

Just within the entrance to the maze there is a small flowerbed. It is pentagonal in shape, as befits a flowerbed in a mathematical maze, and it contains a single row of flowers. There is a lily, a geranium, a delphinium, a marigold, an aster, and three daisies.

As you approach, you disturb a pair of white rabbits, which have been nibbling at the grass verge. One hides behind the geranium, the other behind the delphinium. You are vaguely reminded of a childhood rhyme – but it was about a dormouse☛, was it not? (The symbol ☛ indicates that further information can be found in the 'Pointers' section at the back of this book.)

Flowers, rabbits, a pentagon. A strange selection, presumably serving some symbolic function? You wonder if there is any reason behind it. Idly you count the petals – 'She loves me, she loves me not …'. But, as a good environmentalist, you leave the petals firmly affixed to the flowers. Curious … The lily has three petals, the geranium five, the delphinium eight, the marigold thirteen, the aster twenty-one, and the daisies have thirty-four, fifty-five, and eighty-nine.

Perhaps there is some rationale to the choice of flowers, for the numbers increase steadily. How *steadily? You list the numbers in order:*

$$3, 5, 8, 13, 21, 34, 55, 89.$$

They feel like they ought to have a pattern, but if they do it's not a familiar one. The numbers are not consecutive, like 1, 2, 3, 4, 5, 6, 7. They are not all odd, neither are they all even. They are not powers of two, 2, 4, 8, 16, 32, 64, 128. They are not the primes, 2, 3, 5, 7, 11, 13, 17. They are not squares, 1, 4, 9, 16, 25, 36, 49. One of them, 8, is a cube: $8 = 2 \times 2 \times 2$. The rest are not. Yet they have their own magic, their own aura of significance. They mean something – or else they would not have been planted where they are.

But what? Why have they been planted here, just inside the entrance to the magical maze?

And what do they have to do with rabbits and pentagons?

Passage One

THE MAGIC OF NUMBERS

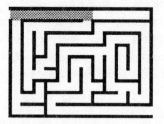

The answer to riddle of the flowerbed is some way off, towards the end of the first passage of the maze – just before we pause for breath, turn through the gap in the wall, and head into the second passage. It is an answer that captures the true magic of numbers, the ability of mathematics to illuminate the secret corners of our world and point to unexpected interconnections. But our first steps must begin with more mundane magic – magic of the stage variety. We'll return to the flowers and the rabbits later, when we're better prepared.

Let's begin with some very simple magic:

- Think of a number.
- Add ten. Double the result. Subtract six.
- Divide by two. Take away the number you first thought of.
- The answer is seven. Always.

We've all come across this kind of trick, and we all know the general principle upon which it rests. Somehow that elusive and unknown number that enters the calculation at the beginning is magically persuaded to disappear again by the end.

Why, though, does it work?

Well-meaning teachers often tell you that the way to understand mathematics is through concrete examples. If you try an example of the party-trick that I've just described, you can easily check that your choice does indeed lead to the result 'seven'. For instance:

Think of a number:	31
Add ten:	41
Double the result:	82
Subtract six:	76
Divide by two:	38
Take away the number you first thought of:	7
The answer is seven: yes	

The trouble is, that calculation gives hardly any insight into why the answer is *always* seven, whatever number you choose.

And that's the first lesson about thinking like a mathematician: sometimes it pays to think in generalities, not specifics. Here, an example doesn't help very much. Of course, you could do a hundred such examples – and I promise you'll always get seven as the answer. The cumulative effect of the examples might convince you that the calculation is rigged so that it always produces the answer seven – but it won't tell you *how* it's rigged. And mathematicians have learned to be very wary indeed of 'experimental evidence', because on plenty of occasions what looked like very strong evidence for some suspected mathematical truth actually turned out to be totally misleading, and the 'truth' was revealed as a falsehood.

If concrete examples don't help, what else can we do?

We can try to think on some level of generality. As simple as possible, but not *too* simple, as Albert Einstein is reputed to have said.

The master-mathematicians of ancient Egypt had a general way of thinking about such questions. They called the unknown number 'heap', meaning 'some fixed but unknown number'. Let's try the calculation with heaps:

Think of a number

Add ten:

Double the result:

Subtract six:

Divide by two:

Take away the heap you first thought of:

The answer is seven: yes.

Easy!

Believe it or not, you've just done two things that most people think of as being difficult and sophisticated.

One is algebra. The other, conceptually deeper, is proof.

Algebra is a kind of symbolic reasoning that works with numbers without knowing their actual values. Proofs provide a once-and-for-all guarantee that certain lines of reasoning always work. Instead of checking lots of examples, you give a logical argument to show that your method *always* works – which in this case means that the answer is always the same number, namely 7. It makes no difference what value you assign to 'heap': after all the arithmetical machinations, the heap magically disappears and only non-heap numbers remain.

Admittedly, our proof, with its little drawings of heaps, doesn't *look* much like algebra. But that's just a matter of notation. To make it look like normal algebra, all you have to do is replace 'heap' by a letter of the alphabet (the traditional one being x) and replace the blobs by conventional numbers. So now the proof looks like this:

Think of a number:	x
Add ten:	$x + 10$
Double the result:	$2x + 20$
Subtract six:	$2x + 14$
Divide by two:	$x + 7$
Take away the number you first thought of:	7
The answer is seven: yes.	

It's the same as the heaps. As Carl Friedrich Gauss, probably the greatest mathematician who ever lived, once said, in mathematics, what matters is *notions*, not notations.

Ideas, not symbols.

Unfortunately, you have to spend a lot of time getting used to the symbols before your mind latches on to the ideas. Still, here we are, only three pages into Passage One, and you've done some algebra and invented a proof.

When you're hot, you're hot!

SNARK ARITHMETIC

Lewis Carroll, most famous as the author of *Alice in Wonderland* and *Through the Looking Glass*, was a mathematician, and he made good use of his mathematics in his writing. His long humorous poem *The Hunting of the Snark* includes a 'think of a number' trick. The Beaver, one of the central characters, has been told that 'what I tell you three times is true'. The Butcher hears a sound. ' 'Tis the voice of the Jubjub!' he declares. ''Tis the note of the Jubjub!' he continues, charging the Beaver with the task of keeping count. ' 'Tis the note of the Jubjub!' he adds. 'The proof is complete, if only I've stated it thrice.' But the poor Beaver goes into a panic at the prospect of encountering the terrible Jubjub bird, and becomes convinced that it must have lost count. The only solution is to *calculate* how many times the Butcher has made his frightening declaration (Figure 1). 'Two added to one – if that could but be done,' it snivels, remembering that in its youth it had 'taken no pains with its sums'. The Butcher, taking pity on the poor creature, affirms that in his opinion the computation is probably possible. At any rate, the Butcher brings paper, ink, portfolio, and pens, and attempts to figure out the answer in a rather unorthodox manner:

> Taking Three as the subject to reason about –
> A convenient number to state –
> We add Seven, and Ten, and then multiply out
> By One Thousand diminished by Eight.
> The result we proceed to divide, as you see,
> By Nine Hundred and Ninety and Two:
> Then subtract Seventeen, and the answer must be
> Exactly and perfectly true.

The procedure is complete nonsense, of course, but that's true of much of the poem. Nevertheless, it is instructive nonsense. If you carry out the arithmetic, you'll find that the 'exactly and perfectly true' answer is three, which is what the Butcher wanted it to be. But there are some clues in the numbers employed which suggest that the Butcher has rigged the calcula-

Figure 1   The Beaver brought paper, portfolio, pens ...

tion so that he can get whatever answer he wants, namely, whichever 'convenient number' he chooses as 'the subject to reason about'.

Among the clues are the occurrence of 'add Seven, and Ten' and, later, 'subtract Seventeen' – instructions that, in the absence of anything in between, would leave the number unchanged. What is in between is an even more blatant clue: 'One Thousand diminished by Eight' is equal to 992, and so is 'Nine Hundred and Ninety and Two'. Since we multiply by the first and promptly divide by the second, again these cancel out.

Circumstantial evidence, but we can do better. Let's work through the algebra, taking x as 'the subject to reason about' – Carroll's term for 'heap'. Here goes:

Taking x as the subject to reason about – x
A convenient number to state –
We add Seven, and Ten, and then $x + 7 + 10 = x + 17$
multiply out
By One Thousand diminished by Eight. $(x + 17) \times (1000 - 8) = (x + 17) \times 992 =$
$992x + 16,684$ ☜
The result we proceed to divide, as you see,
By Nine Hundred and Ninety and Two: $(992x + 16,864)/992 = x + 17$
Then subtract Seventeen, and the
answer must be $(x + 17) - 17 = x$
Exactly and perfectly true x

Just as we suspected!

THE MYSTERIOUS NINE

With a little ingenuity, this kind of self-fulfilling process can be disguised in all sorts of ways. There is a stage magician's trick which exploits the same general principle in a geometric guise. The magician places twenty or so coins on the table in the form of a figure 9 (Figure 2). The 'tail' of the 9 contains, say, seven coins; the rest form a closed circle. The tail runs into the circle at a coin which I shall call the 'junction'. While the magician's back is turned, a victim selected from the audience is told to think of a number bigger than seven (the number of coins in the tail). They start counting from the tip of the tail, running anticlockwise round the circle, until they reach their

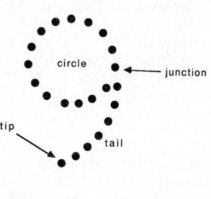

Figure 2   The mysterious nine.

chosen number. Then they start counting at 1 again, moving one coin clockwise for each number in the count, and stop on reaching their chosen number. The victim conceals a tiny piece of paper with the message THIS ONE underneath this coin.

The magician now turns round and unhesitatingly picks up the same coin.

How does he do it? In fact, the counting procedure will always end on the same coin, no matter what number was chosen – as long as it is bigger than the number of coins in the tail of the 9. The part of the anticlockwise count that goes round the circle is cancelled out by the corresponding part of the clockwise count. If the second count were to divert back down the tail, then of course it would end at the tip – exactly where it started. The victim would have counted the chosen number twice, the second count exactly undoing the first because the direction is reversed. However, the magician has cunningly insisted that the second count should stay on the circle. So the victim arrives at the junction with seven counts left to go – the number of coins in the tail – and so his second count ends on the seventh coin clockwise from the junction.

Here it is in algebra. Suppose the victim chooses x, bigger than 7. He counts 7 coins up to the junction, which leaves an anticlockwise count of $x-7$ from the junction onwards. This is followed by a further x in the clockwise direction. The first $x-7$ return the count to the junction, so that $x-(x-7)=7$ coins remain to be counted. Therefore the count ends on the seventh coin clockwise from the junction.

This time an example does illuminate the argument (teachers are often right, too). Figure 3(a) shows how the count goes if the victim chooses the number 13. The trick is especially transparent if the victim makes the second count backwards, from 13 to 1, as in Figure 3(b). Figure 3(c) does it with x's.

The trick can be repeated, but to stop the audience realising that the count always ends on the same coin, no matter what the victim chooses, you should use different numbers of coins and vary the number in the tail.

TAP-AN-ANIMAL

There are other ways to make a chosen number 'cancel out'. Sometimes the number itself can be disguised, so that the audience is less inclined to look for arithmetical trickery. A delightfully simple example of this is the Tap-an-Animal trick, devised by Martin Gardner. Gardner is a journalist

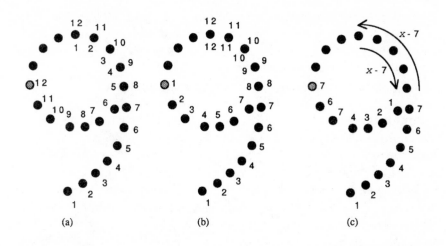

Figure 3 How the mysterious nine works: (a) an example; (b) the same example with backwards numbering; (c) doing it with x's.

and writer who for many years wrote a regular Mathematical Games column in *Scientific American*; he is also an amateur magician. In this trick, intended for young children, his two interests come together.

The trick makes use of a diagram featuring eight animals at the tips of an eight-pointed star (Figure 4). The victim chooses an animal – say the monkey. The magician starts tapping his way round the star with his wand, beginning at the butterfly, going on to the rhinoceros, and continuing in the same direction, anticlockwise round the star. The victim silently spells out the name of the chosen animal, one letter per tap – M, O, N, K, E, Y – and shouts 'stop!' upon reaching the final letter.

Lo and behold, the wand rests on the monkey.

The secret is straightforward. Three taps round the star is the COW (three letters). Four taps round is the LION (four letters). Five taps round is the HORSE (five letters). In general, an animal with x letters in its name lives x taps round the star. Starting with BUTTERFLY is a device to distract attention from this pattern – BUTTERFLY has nine letters, and since the star has eight points, the ninth tap returns to the position of the first. And RHINOCEROS, with ten letters, lives on the second point of the star – which is also reached after ten taps, because $10 = 8 + 2$.

Adults quickly grasp the basis of the trick, but similar games can be played which puzzle even adults. Here's an unusual example, which involves a board with numbers and letters (Figure 5). Get your victim to choose any number on the board, and then spell it out, letter by letter.

Figure 4　Tap-an-Animal.

Add together the corresponding numbers (subtracting those on black squares, adding those on white squares). The result will always be plus or minus the number you chose. For instance, if the victim chooses 11 you spell out ELEVEN, which translates into $-4+24-4+1-4-2$, which is 11. Lee Sallows, the inventor, had to do a lot of clever mathematics to make that trick work.

In 1940 Gardner produced an advertising free gift, the Magic Tap-a-Drink card, based on the same principle. The card had ten holes, arranged in a circle, and beside each was the name of a cocktail. The victim picks a cocktail, then the magician turns the card over and taps a pencil round the holes in turn while the victim silently spells out the drink. Then the magician pokes the pencil through the hole, and turns the card over, to show that he has correctly located the cocktail. Figure 6 shows a variant

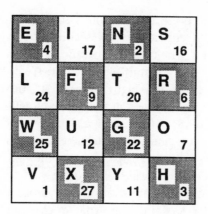

Figure 5   Lee Sallows' magic square.

that uses numbers instead. Start at the arrow and tap in a clockwise direction (ignore hyphens when spelling the names). SIX, with three letters, is in position three, and so on. For better effect, use the number symbols, not the names: I've written the names on the diagram to make the principle clearer.

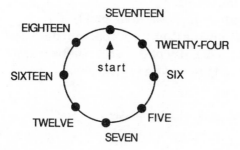

Figure 6   Tap-a-Number.

DAYS IN THE WILDERNESS

A certain amount of clever concealment goes into these puzzles. Although the number in position three (namely 6) has three letters, the number in position two (24) doesn't have two letters. Instead, it has ten letters. This makes the numerical pattern less obvious. Why do we use ten here? Because most other numbers won't work. If you count ten steps round a circle of eight dots, then you arrive at the second dot anyway.

You wouldn't have done that if you'd counted nine dots, or eleven. However, you'll still end up on the second dot if you count 18 steps round, or 26, or 34 … The common feature of these numbers is that they are all two greater than a multiple of eight: $0+2$, $8+2$, $16+2$, $24+2$, $32+2$. Since the circle has eight dots in it, moving round it eight dots brings you back to where you started. So, in effect, 8 is 'the same as' 0. Every multiple of 8 is 'the same as' 0, for much the same reason: you may go round more times, but you still end up where you started. What determines where you end up is not the number of dots you count, but how many of them are left after you've got rid of multiples of eight.

Buried in this innocent observation is a profound variation on the traditional number system, one that proves invaluable whenever numbers 'wrap round' to eat their own tails. This event is far more common than my description might suggest. It happens, for example, in many areas of applied science: those where the same sequence of events repeats indefinitely. Astronomical cycles, such as the motion of the Moon round the Earth or the Earth round the Sun, are typical examples.

Human calendars – I use the plural for good reasons, as you'll shortly see – are based on astronomical cycles. Not surprisingly, this kind of 'wraparound arithmetic' is very useful in setting up and comparing calendar systems.

Let's start with a simple warm-up problem.

It is Sunday. John the Baptist goes out into the wilderness, due to return after forty days. What day of the week will he return?

Well, you could just make a big list (Table 1).

Table 1

Day of the week	Number of days					
Sunday	0	7	14	21	28	35
Monday	1	8	15	22	29	36
Tuesday	2	9	16	23	30	37
Wednesday	3	10	17	24	31	38
Thursday	4	11	18	25	32	39
Friday	5	12	19	26	33	40
Saturday	6	13	20	27	34	

The list tells you that John will reappear on Friday. But it's a cumbersome method. What if John had gone into the wilderness for forty thousand days? Or forty million? We can't just make bigger and bigger lists!

Fortunately, there's a pattern in the list, and once you've seen it you don't need the list at all. Let's take a look at the Sundays. These occur on days 0, 7, 14, 21, 28, and 35 of John the Baptist's sojourn. It's not hard to see the pattern: these are the multiples of 7. OK, now what about Mondays? Those occur on days 1, 8, 15, 22, 29, 36. Again, there's a pattern – but the easiest way to see it is to observe that Monday is always one day later than Sunday. So Mondays correspond to days that are 1 more than a multiple of 7. By the same argument, Tuesdays correspond to days that are 2 more than a multiple of 7, and so on.

How can we tell that a number is, say, 2 more than a multiple of 7? Easy: divide by 7 and see if the remainder is 2. For example, 72 divided by 7 goes 10 times with remainder 2 – meaning that $72 = 10 \times 7 + 2$ – so day 72 would be a Tuesday as well. So would day 702, or day 7,000,002.

In other words, what we should do is give the days of the week code numbers, as in Table 2.

Table 2

Day of the week	Code
Sunday	0
Monday	1
Tuesday	2
Wednesday	3
Thursday	4
Friday	5
Saturday	6

Then the day of the week for a given number of days in the wilderness is whichever day corresponds to the *remainder* on dividing that number by 7.

For 40 days' sojourn, all we do is divide 40 by 7 and find the remainder. Now, 7 into 40 goes 5 times, with remainder 5 (that is, $40 = 7 \times 5 + 5$). So the answer is whichever day had code 5, and that's Friday.

For 40,000 days' sojourn, we divide 40,000 by 7 and find the remainder. Now, 7 into 40,000 goes 5714 times, with remainder 2 (that is, $40,000 = 7 \times 5714 + 2$). The day whose code is 2 is Tuesday.

For 40,000,000 days' sojourn, we divide 40,000,000 by 7 and find the remainder. Now, 7 into 40,000,000 goes 5,714,285 times, with remainder 5 (that is, $40,000,000 = 7 \times 5,714,285 + 5$). The day whose code is 5 is Friday, again. If John had stayed for forty million days, he would have come out on the same day of the week that he did after forty days.

MODULAR ARITHMETIC

Mathematicians have formalised this kind of calculation. The idea was first systematised by Carl Friedrich Gauss in his monumental work *Disquisitiones arithmeticae*, the second great number theory book☞. He called it 'arithmetic to a modulus', and it goes like this. Choose a whole number, and for reference call it the *modulus*. In John-the-Baptist arithmetic, the modulus is 7, and for the sake of illustration we'll use that modulus for the moment.

Say that two numbers x and y are *congruent* to the modulus 7, written

$$x \equiv y \quad (\mathrm{mod}\, 7)$$

if their difference x-y is an exact multiple of 7. For example,

$$40 \equiv 5 \quad (\mathrm{mod}\, 7)$$

because $40 - 5 = 35 = 5 \times 7$.

The solution to the John-the-Baptist puzzle boils down to this: day numbers that are congruent mod 7 have the same code, and thus correspond to the same day of the week. Day 40 has the same code as day 5, and is therefore a Friday.

In fact, we can quickly find the code from the day number by observing that it is equal to the remainder on dividing the day number by 7. The remainder is what's left over when you subtract the biggest multiple of 7 that does not exceed the day number, and as such it must always be one of 0, 1, 2, 3, 4, 5, or 6.

Gauss observed that you can do arithmetic mod 7, using *only* the seven numbers 0, 1, 2, 3, 4, 5, and 6 – and that the usual algebraic rules still apply. Well, there's *one* rule that no longer applies: the assumption that 7 is different from 0. In arithmetic mod 7, the number 7 is the same as 0 – well, it's congruent to 0, which is effectively the same thing.

For example, you can add numbers mod 7 by adding them in the usual way and then replacing the result by its remainder on division by 7. So

$$2 + 3 \equiv 5 \quad (\mathrm{mod}\, 7),$$
$$3 + 5 \equiv 8 \equiv 1 \quad (\mathrm{mod}\, 7),$$
$$4 + 6 \equiv 10 \equiv 3 \quad (\mathrm{mod}\, 7),$$

and so on. You can also multiply two numbers (mod 7). Again, you do this by multiplying in the usual way and then replacing the result by its remainder on division by 7. For example,

$$3 \times 6 \equiv 18 \equiv 4 \quad (\mathrm{mod}\, 7),$$
$$4 \times 5 \equiv 20 \equiv 6 \quad (\mathrm{mod}\, 7).$$

To illustrate how powerful this idea is, we can use arithmetic mod 7 to calculate on what day of the week John the Baptist returns after 40,000 days in the wilderness. I'll omit the (mod 7) symbol – but bear it in mind. Here goes:

$$40,000 \equiv 4 \times 10 \times 10 \times 10 \times 10$$
$$\equiv 4 \times 3 \times 3 \times 3 \times 3 \equiv 4 \times 9 \times 9$$
$$\equiv 4 \times 2 \times 2 \equiv 4 \times 4 \equiv 16 \equiv 2.$$

So John comes home on Tuesday – just as we found before.

PREDICTING THE DATE OF NEW YEAR'S DAY

Now let's get more ambitious, and look at the whole calendar – not just forty days of it. The same principles that made it easy to find out the day of John's reappearance explain all sorts of strange things in calendars. The subject is called chronology – the science of the measurement of time.

Cosmologists take an inordinate interest in the first few microseconds after the Big Bang, but pay hardly any attention to what happened thereafter. Chronologists would be interested in the Big Bang only if you could tell them what date it happened on, but they will tell you that what happened thereafter was so inordinately complicated that you need degrees in mathematics, astronomy, sociology, divinity, and law even to begin to comprehend it. Take, for example, the simple question of the date of New Year's Day. Up until now you may well not have realised that there *is* a question here, but by the time I've finished you probably won't even be able to remember your own birthday.

New Year's Day is 1 January, of course.

Well, yes – in countries that have adopted the Gregorian calendar and after the time they adopted it. Which in Italy and France was 1582, in England 1752, in Germany between 1583 and 1700 depending on province, and in Finland as late as 1918. (Throughout, all dates will be AD unless specified as BC.) However, the Italian provinces of Treviso, Tuscany, and Venice changed to the Gregorian calendar but retained 25 March for New Year's Day until 1750, and ...

I'm oversimplifying, of course. Sorry about that.

Prior to the adoption of the Gregorian calendar, a variety of dates for the new year were favoured by different states at different times and for different purposes. 1 January was adopted by the Roman Empire under Julius Caesar. Until 800 France preferred 1 March; then they switched to 25 March until 996, after which New Year's Day coincided with Easter until 1051 … Between the seventh century and 1338 the English considered Christmas Day to be the start of the new year, but in 1339 they moved it to 25 March for civil purposes and Easter for religious purposes.

Of course, we've got it all sorted out nowadays? Sorry, no, not so as a visiting alien would notice. Even in Britain we have to decide whether we're using the calendar year (1 January), the financial (6 April), or the academic year (1 October). You may feel that 1 January is the only one that counts as the *real* New Year's Day. I agree, but in 1998, for instance, New Year's Day will not fall on 1 January everywhere in the world. According to the Chinese calendar it will be on 28 January, for the Burmese 16 April, in the Islamic world 28 April, and for Jews 21 September☞.

Though for them it won't actually be the year 1998, you understand.

The history of the calendar – sorry, calendar*s*, by the bucketful – is a long-running planet-wide soap opera, sweeping from glorious highs like Julius Caesar's addition of 90 days to the year 46 BC to get the seasons back into their rightful places, to atrocious lows such as the muddle that the Roman priests made of his rule for allocating leap years, which was not fully corrected until AD 4. It is a wonderful example of humanity's most endearing and infuriating trait; the inability to get the simplest and most basic things right. Or even consistent.

CURIOUSLY CYCLING COSMOS

In humanity's defence, it must be said that the cosmic dice are loaded against us. Our chronic calendric confusion has its seeds in what Isaac Newton called 'the System of the World', the movement of the bodies that make up the solar system. So before we can turn to the tricky task of predicting the date of the new year, we must first review some fundamental facts about the heavens as perceived from the Earth.

The Earth rotates once on its axis every 23.9345 hours, or 23 hours 56 minutes and 4 seconds. This is the time it takes for one rotation relative to the 'fixed stars' (which of course aren't fixed, but never mind that). The period known as a day differs from this, because a day is the period between successive occasions when the Sun is overhead. Now, while the Earth is rotating about its own axis, it is also revolving around the Sun, and it takes a further four minutes or so for the extra rotation to compensate for the Sun's apparent slippage back across the sky. This leads to the more familiar figure of 24 hours in a day.

To ancient humanity, the next most obvious cycle in the sky was the sequence of phases of the Moon. Relative to the fixed stars, the Moon revolves around the Earth once every 27.32166 days. However, its phases are governed by the relative positions of the Sun and the Moon as observed from the Earth, and once more there is some slippage in the Sun's position which has to be made up, leading to a 'synodic lunar month' of 29.53059 days.

Finally, there is the year – the time that it takes the Earth to go once round the Sun. Relative to the fixed stars, this is the 'sidereal year' of 365.25636 days. For calendric purposes, however, the more important period is the time between successive spring equinoxes, which is the 'tropical year' of 365.24219 days.

The important point here is not so much the numbers themselves, or the complicated dynamic geometry that gives rise to them. It is just that neither the tropical year, nor the lunar month, are simple multiples of a day. This implies that any calendar that employs periods of a day and attempts to fit them either to the motion of the Sun or to that of the Moon – let alone both – will not be able to follow a fixed periodic cycle.

The mathematical distinction here is that between *rational* numbers, which are fractions like 3/4 or 355/113 made up from two whole numbers, and *irrational* numbers, such as $\sqrt{2}$ or π (pi), which are not representable as exact fractions. If the solar year were a rational multiple of a day, then after some integer number of days, the year and the day would be back in step at exactly the same point. For example, if the tropical year were exactly 365.25 days, a ratio of 1461/4, then 4 years would be exactly equal to 1461 days, and all astronomical events related to the apparent position of the Sun would repeat precisely over that period. However, 4 years *isn't* exactly equal to 1461 days, and although any number can be approximated by a fraction, the numbers involved get too big to be practical. So for practical purposes, both the tropical year and the lunar month contain an irrational number of days. Worse, the tropical year also contains an irrational number of lunar months. So *nothing ever repeats exactly.*

This is curiously different from our earlier excursions into repetitive cycles, such as the eight dots round the circle in the Tap-an-Animal trick, or the seven days of the week. There, we could find a place where everything began to repeat exactly. The difference now is that we can make one cycle repeat exactly, or another, but not both at once – not with the same period.

Imagine the seven days of the week wrapped round a circle as a series of seven equally spaced dots. Day seven is the same as day zero: we have come full circle. We can fit hours into the same cycle: twenty-four hours make one day, so all we have to do is subdivide each day into 24 equal arcs. Now 24 hours is different from 0 hours – it moves us from one day to the next. But $7 \times 24 = 168$ hours fits perfectly into the circle – it takes us from (say) 3 o'clock this Sunday to 3 o'clock next Sunday. There is no problem, as the years roll by, of the hours gradually slipping out of synchrony with the days.

That is not true, however, of the month – or the year. A month of 27.32166 days wraps just less than four times round the circle, falling short by 0.67834 days. By the second month, this slippage has grown to twice the size: 1.35668 days. The slippage continues, following the same pattern, until after ten months it has grown to 6.7834 days. However, this is just short of a week, and so is equivalent to a slippage of 0.2166 days (equal to 7 - 6.7834). So suddenly the weeks and months seem to have become a lot closer to synchrony than they were at the start. Appearances are deceptive, because this discrepancy itself grows ... Only after a huge number of months have passed does the slippage relative to the weeks become zero – and then only because we have stopped measuring the length of the month at some fixed number of decimal places. To be precise, 100,000 months equals 2,732,166 days, or 390,3094/7 weeks. So 700,000 months equals 2,732,166 weeks. This equality would be exact if a month were really precisely 27.32166 days – but the true figure is probably something like 27.321664, say. In this case, slippage continues: the end of a week *never* coincides with the end of a month – not exactly.

Calendars therefore have to make compromises. It is the history of those compromises, in different cultures, that has led to today's plethora of calendric systems, and the existence of at least twenty-six different New Year's Days in any given year – an average of one a fortnight.

JULIUS'LL FIX IT

There are two basic types of calendar: lunar and solar, in which primary attention is paid to the apparent motion of the Moon and Sun, respectively. Most lunar calendars include clever solar-related jiggles to keep them roughly in tune with the seasons; most solar calendars at least pay lip-service to the movements of the Moon.

Our own calendar is solar, and goes back to ancient Rome. In 46 BC Julius Caesar reformed the previously erratic Roman calendar, a lunar one. On the advice of the astronomer Sosigenes he took the length of the tropical year to be 365¼ days, and set up a cycle of 1461 days consisting of one leap year of 366 days and three common years of 365. There were twelve months, relics of the lunar system but no longer linked to the lunar month. Their lengths were not quite those we have now; in particular, February had 29 days in a common year and 30 in a leap year.

Our current lengths seem to have been introduced by Julius's successor Augustus Caesar, and an apocryphal story has it that he pinched a day from February to make August (renamed after him from its original Sextilis) the same length as July (previously renamed from Quintilis in honour of Julius).

Around 10 BC it was discovered that the priests had followed the common Roman procedure of beginning a new count with the end of the previous one, thereby creating a leap year every *third* year by mistake, so leap years were omitted altogether until AD 4 to get the calendar back on track. Julius added 90 days to the year 46 BC – a standard 'intercalated' month from the old lunar calendar, plus two new ones to get spring back to its traditional date of mid-March. Julius decreed that the year would start on 1 January, proving what a sensible chap he was.

In actual fact, a Julian year exceeds the true tropical year by 0.00781 days, so by 1582 the spring equinox had slipped back from 21 March to 11 March. To prevent further slippage of the seasons relative to the calendar year, Pope Gregory XIII reformed the calendar once more. From that point on, a year ending in '00' would be a leap year only if it was a multiple of 400. This resulted in the omission of 3 days from every 400-year cycle, and reduced the 'theoretical' length assumed by the structure of the calendar to 365.2425 days, much closer to the true value of 365.24219. To bring the equinoxes back into alignment, ten days of 1582 were removed, 5 October becoming 15 October. The new year, as in the Julian calendar, began on 1 January☞.

In the Julian calendar, as it happens, 1 January of year 1 was a Sunday. What date occurs one million days after 1 January of year 1 in the Julian calendar?

The day of the week will be 1,000,000 (mod 7) = 1 day after Sunday – namely Monday. But what of the year, month, and day? Let's start with a simpler warm-up problem. If every year were exactly 365 days long – a system that I'll call the *Sosigenean* calendar – this would be easy. Everything would repeat after 365 days. Divide 1,000,000 by 365 and find both quotient and remainder.

$$1,000,000 = 2739 \times 365 + 265.$$

This tells us that one million days is the same as 2739 Sosigenean years and 265 days. So the date would be the 265th day of the year AD 2740 (Sosigenean). Note that we have to add 1 because we started the count in year 1, not year 0. In our current Gregorian calendar, this is why the third millennium really starts on 1 January 2001, not 1 January 2000.

A further calculation shows that this date would be 22 August, assuming the usual system of months but with no leap year day of 29 February. Lengths of months being rather messy, there is no especially neat mathematics involved in this step in the calculation, but here's the idea. The end of August occurs 243 days into the year. That leaves 22 to go, so the date is 22 August.

What are we doing here? Let's use the notation.

$$m \text{ div } n = \text{quotient on dividing } m \text{ by } n$$

as well as our previous

$$m \text{ mod } n = \text{remainder on dividing } m \text{ by } n.$$

Then the number of *complete* years that passes in 1,000,000 days is

$$1{,}000{,}000 \text{ div } 365,$$

and the number of days left after that is

$$1{,}000{,}000 \text{ mod } 365.$$

The same kind of calculation applies with 1,000,000 replaced by any other number. In short, the date of day N of the Sosigenean calendar would be

$$\text{year } N \text{ div } 365 + 1, \quad \text{day } N \text{ mod } 365,$$

where the +1 arises, as before, because we started in year 1.

ONCE AGAIN, WITH LEAP YEARS ...

In the true Julian calendar, however, some years have 366 days – namely, those years that are multiples of 4. This changes the answer.

One way to correct the calculation is to start with cycles of four years, like this:

years				days
1	5	9	and so on have	365
2	6	10		365
3	7	11		365
4	8	12		366.

Each four-year cycle contains $365 + 365 + 365 + 366 = 1461$ days. So the number of complete cycles of this kind is

$$1,000,000 \text{ div } 1461 = 684 = C, \text{ say,}$$

which accounts for $4C$ years, and the number of days left over is

$$1,000,000 \text{ mod } 1461 = 676 = R, \text{ say.}$$

Those R days consist of

$$R \text{ div } 365 = 1 = D$$

complete years (because the first three years of the cycle all have 365 days, and we don't include all of the fourth year since we have eliminated any complete four-year cycles). The number of days left over is

$$R \text{ mod } 365 = 311 = S, \text{ say.}$$

The total number of years that have passed is $4C + D = 2737$, and we want the 311th day of the corresponding year. Again, this means that the actual year is 2738 – add 1 for the start year. That's not a leap year, so its 311th day is 7 October.

Again, the same kind of calculation will work for any number of days. And a more elaborate version will handle the Gregorian year. Software is now available, on disk or on the Internet, for converting dates between almost any calendric systems☞.

FIBONACCI'S RABBITS

Whoops, I nearly forgot about the rabbits.

You'll recall that just inside the entrance to the magical maze we noticed a pentagonal flowerbed, infested by rabbits, containing a series of flowers whose petals were numbered

$$3, 5, 8, 13, 21, 34, 55, 89.$$

These strange numbers are called the Fibonacci numbers, and they have a lengthy history. They were invented by Leonardo of Pisa in the year 1202. Leonardo was (much!) later☞ given the nickname 'Fibonacci' – based on the Italian for 'son of Bonaccio'. His most important contribution to humanity was to popularise – among Europeans – the newfangled Hindu-Arabic 'place notation'. Essentially this is our current decimal notation for numbers, in which the same symbol has different values depending on where it is. For example, in the Fibonacci number 55, the first '5' signifies 'fifty' and the second one 'five'.

Fibonacci introduced his numbers in a problem about rabbits. (I'll get to the flowers later.) His rabbits are immortal, arrange themselves in pairs, and breed once per season. In the beginning there is one pair of immature rabbits. These mature for one season. The next season they give birth to another immature pair. The same pattern persists for ever: each mature pair gives birth to one immature pair every season; each immature pair takes one season to mature.

How many rabbits are there after a given number of seasons? One way to answer Leonardo's question is to build up a short table, like Table 3.

Table 3

season	immature	mature	total
1	1	0	1
2	0	1	1
3	1	1	2
4	1	2	3
5	2	3	**5**
6	3	**5**	8
7	**5**	8	13
8	8	13	21

In order to get from one row of the table to the next:

- Copy the number of mature pairs into the 'immature' column.

- Add the numbers of mature and immature pairs together and write that in the 'mature' column.

- Add the two and put the result in the 'total' column.

Apart from the first few lines, any given number appears in the table precisely three times – for example, 5 appears where I've marked it in bold type. On one row it is in the 'total' column, on the next it is in the 'mature' column, and on the one after that it is in the 'immature' column. It is easy to see why this pattern occurs, and that it will persist for ever:

- This year's total is next year's number of mature pairs, since all current immature pairs will mature, and the mature ones remain alive.
- This year's number of mature pairs will become next year's immature pairs, because each mature pair breeds one new immature pair.

Now, look at the final row in the short table. Its total, 21, is obtained by adding the 13 and 8 in the same row. But the 13 also appears as a total in the previous row, and the 8 appears as a total in the row before that. In other words:

- *This year's total is the sum of last year's and the year before's.*

This is the hidden rule behind the Fibonacci numbers. To generate them all, start with the short sequence

$$1, 1,$$

and then form a new number by adding them together:

$$1, 1, 2 \quad (=1+1).$$

Then add the last two numbers there to get

$$1, 1, 2, 3 \quad (=2+1),$$

and repeat:

$$1, 1, 2, 3, 5 \quad (=3+2),$$
$$1, 1, 2, 3, 5, 8 \quad (=5+3),$$

and so on.

GOLDEN RATIO

We've found a *rule* for generating the Fibonacci numbers, but not – yet – a formula. As it happens, there is a simple formula, and we can work our way towards it by looking at the ratios of consecutive Fibonacci numbers. Like many populations whose growth is unrestricted, such as bacteria newly introduced into a biologist's Petri dish, Fibonacci's rabbits grow exponentially – that is, each successive population is (almost) a fixed multiple of the previous population. The ratio gives the growth rate. So of course the alert mathematician should look at ratios.

Let's see what happens:

$$1/1 = 1.000,$$
$$2/1 = 2.000,$$
$$3/2 = 1.500,$$
$$5/3 = 1.666,$$
$$8/5 = 1.600,$$

$$13/8 = 1.625,$$
$$21/13 = 1.615,$$
$$34/21 = 1.619,$$
$$55/34 = 1.617,$$
$$89/55 = 1.618 \ldots$$

The ratios seem to be getting closer and closer to the same value, something around 1.618. In fact, if you keep generating bigger and bigger Fibonacci numbers, it turns out that the ratio gets as close as you wish to the so-called *golden number*

$$\phi = (1 + \sqrt{5})/2 = 1.618034 \ldots.$$

The symbol ϕ is the Greek letter 'phi'. The golden number has the very curious property that

$$1/\phi = 0.618034 \ldots = \phi - 1$$

or

$$\phi = 1 + 1/\phi$$

To see the connection between the golden number and the Fibonacci sequence, start with an equation like

$$89 = 55 + 34$$

and divide out by 55, to get

$$89/55 = 55/55 + 34/55.$$

Suppose that the ratios of consecutive Fibonacci numbers are all very close to some particular value, call it x. Then 89/55 is approximately x, and 34/55 is very close to $1/x$ (since it is 55/34 upside-down). Of course, $55/55 = 1$. So, to a close approximation,

$$x = 1 + 1/x,$$

and this equation is satisfied when $x = \phi$, the golden number.

None of this *proves* that the ratios get close to ϕ, but it does show that if they get close to anything, ϕ is pretty much the only possibility. With a bit of algebraic technique, the whole story can be made logically sound. The upshot is that each Fibonacci number is very nearly equal to ϕ times the previous one. We therefore expect the Fibonacci series to be (very nearly) *proportional* to the powers of ϕ – namely, 1, ϕ, ϕ^2, ϕ^3, ϕ^4, and so on. In fact, the Fibonacci numbers are precisely the nearest whole numbers to

$\phi^n/\sqrt{5}$, for $n = 1$, 2, 3, ...☞. This is the promised formula for Fibonacci numbers, and it shows how close they come to exact exponential growth.

The ancient Greeks were fascinated by the golden number, because it is central to the geometry of the regular five-sided polygon, or *pentagon*. Suppose you draw a perfect pentagon with sides one unit long (Figure 7(a)). By joining each vertex to the next but one you get a five-pointed star (Figure 7(b)). The Greeks could prove that the sides of the star have length ϕ, and this led them to a geometric construction for a pentagon using only ruler and compasses.

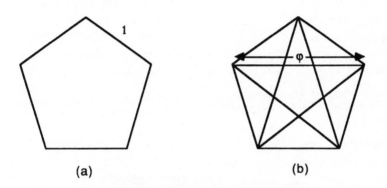

(a)   (b)

Figure 7   (a) Pentagon with unit sides; (b) sides of inscribed star have length ϕ.

FLOWER PATTERNS

We now know why the flowerbed by the entrance to the maze is infested by rabbits, why it is pentagonal, and why the petals on the flowers are Fibonacci numbers. The maze-builder (yours truly) has put them there because all these things are related by the same mathematical package: properties of the golden number.

Why flowers, though?

It has been known for several hundred years that *most* flowers have a Fibonacci number of petals. You find 55 petals in many daisies, for instance, but you hardly ever find 56 or 54 (unless, in the latter case, one has fallen off). When you don't find Fibonacci numbers, you usually find numbers twice as big:

$$2, 4, 6, 10, 16, 26, 42,$$

or numbers in the 'anomalous series'

$$4, 7, 11, 18, 29, 47,$$

which are formed by applying the Fibonacci rule but starting with 4 and 7, not 2 and 3.

Why is the plant world obsessed with Fibonacci numbers?

We now have a fairly complete answer; to get it has taken only a couple of centuries. The first step was to see the problem not as one in number theory, but as one in geometry. The heads of many plants bear seeds in beautiful swirling spiral patterns. These patterns are especially striking in ripe sunflowers (Figure 8), but they also occur in many other species, notably daisies. To the human eye, those patterns consist of two families of spirals: one swirling clockwise, the other anticlockwise. If you count how many spirals there are in each family, you get answers like '21 and 34' or '34 and 55'.

Consecutive Fibonacci numbers.

Why?

The interesting bits of flowers – petals, seeds, sepals, stamens, whatever

Figure 8   Spiral patterns in sunflowers.

– all grow from small lumps of plant tissue called primordia. As the plant shoot grows, new primordia appear at the tip and move outwards. The scientific brothers Auguste and Louis Bravais discovered that new primordia appear in predictable places. In fact, the angle between successive primordia is generally very close to 137.5º. They didn't know why, but their experiments demonstrated that it was so. The same angle occurs in the seed-heads of the sunflower: seeds that grow from consecutively formed primordia are separated by an angle of 137.5º. The main problem, then, is to relate the geometry of that angle to the Fibonacci numbers; once this is done, the rest of the numerology, such as that of petals, follows fairly easily. You can argue that petals develop from primordia at the outer ends of one family of spirals, for instance.

How does 137.5º relate to Fibonacci? The connection is surprisingly direct, once you see what it is.

You can measure any angle in two ways: internally or externally. If the internal measure is 137.5º, then the external one is 360º – 137.5º, which is 222.5º. If you form the ratio 360/222.5 you get 1.618 – suspiciously similar to the golden number ϕ. Accordingly, we call 137.5º the *golden angle*. It is equal to $1/\phi$ of a full turn (360º), measured externally.

We can now see why Fibonacci numbers turn up in the spirals. Suppose for a moment that instead of the angle between successive primordia being $1/\phi$, it was a ratio of consecutive Fibonacci numbers such as 5/8. (Since ϕ is close to 8/5, it follows that $1/\phi$ is close to 5/8.) What arrangement would you get by forming successive primordia at angles separated by 5/8 of a full turn?

You'd get the arrangement of Figure 9 – eight radial 'spokes'. The successive angles would be the following multiples of one full turn:

$$0,$$
$$5/8,$$
$$10/8 = 2/8,$$
$$15/8 = 7/8,$$
$$20/8 = 4/8,$$
$$25/8 = 1/8,$$
$$30/8 = 6/8,$$
$$35/8 = 3/8,$$
$$40/8 = 0,$$

and the angles would repeat thereafter.

Here we discard any angle that amounts to one full turn. For instance, 10/8 is 8/8 + 2/8, and 8/8 = 1 is one full turn. Another way to say this is

Figure 9 If successive primordia were to appear at intervals of 5/8 of a turn, they would be arranged in eight 'spokes'.

that we reduce the numerator to its value mod 8. So the successive angles lie at positions

$$(5n \bmod 8)/8$$

for $n = 0$, 1, 2, 3, ... Notice that we get eight radial spokes here because we have chosen an angle 5/8, which is a *rational* multiple of a full turn – an exact fraction. If instead we had used 8/13, we would have got 13 radial spokes. The true golden angle lies between these two values. It is slightly smaller than 5/8 and slightly greater than 8/13. So neither picture of perfect radial spokes is quite right as a description of primordia placed according to the golden angle.

If we use an angle *very* slightly smaller than 5/8, the picture of radial spokes is still pretty good. But, like the seasons relative to the Julian year, there is a small degree of drift. As we go round and round the circle, putting in primordia, those at the tips of the spokes don't quite coincide with their predecessors. The slippage makes the spokes bend slightly. We still see eight spokes, but now they curve (in a clockwise direction, assuming that the sequence of primordia goes anticlockwise). Much the same goes for 8/13, but now we make the angle slightly bigger, getting 13 curved spokes, curving in the anticlockwise direction.

At the golden angle, *both* these pictures are valid simultaneously. Our eye can align the primordia in either 8 clockwise spirals or 13 anticlockwise ones.

That's why we see Fibonacci numbers of spirals. Ratios of Fibonacci numbers provide the best approximations to the golden angle. If those approximations were exact, we would see radial spokes with either one Fibonacci number or the other. Because the approximation is not exact, we see a compromise in which we can pick out one family of curved spikes (spirals) for one Fibonacci number, and another family for the next Fibonacci number.

Of course, by arguing the same way with (say) 34/55 and 55/89, I could explain why we see 55 spirals one way and 89 the other – which conflicts with my explanation of why the numbers should be 8 and 13. The actual numbers that arise depend on how big the seed-head is, and slight variations in the rate at which the primordia migrate away from the tip of the growing shoot. But I think you can see that once we have the golden number, then Fibonacci numerology cannot be far away. The rest is mere detail.

There is a good reason *why* nature chooses the golden angle. A plant whose seed-head consisted of radial spokes would not be very strong – the spokes would be long and thin and fragile. H. Vogel noticed that when you use the golden angle, the primordia pack together very efficiently, with hardly any missing space (Figure 10). This gives a nice, solid, compact seed-head. Now, number theorists know that of all irrational numbers, the golden number is the one that can be approximated most poorly by rational numbers. (That is, for a given size of integers m and n, the difference between m/n and ϕ is generally worse than for any other irrational☞.) So the golden angle is the one that gets as far away as possible from forming radial spokes. It is hardly surprising that it goes to the other extreme: dense packing.

More recent work still, by the French physicists Yves Couder and Stefan Douady, has shown that this choice of angle is a natural consequence of the dynamics of a growing plant shoot☞. Each new primordium gets pushed into the biggest space available. That means that they all pack together efficiently, and that in turn implies that the golden angle is the most likely choice. Dramatically, it turns out that the next best alternative leads to flower numerology from the 'anomalous sequence' 4, 7, 11, 18, 29, 47, ..., which was widely known to be the next most common sequence of numbers chosen by flowers. So the theory explains the exceptions, as well as the more common Fibonacci numbers.

BEFORE WE TURN THE CORNER ...

We have solved the riddle of the rabbit-infested flowerbed. Now we

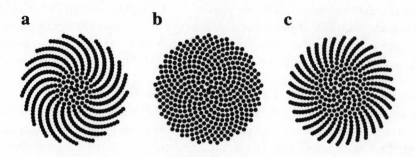

Figure 10   Packing primordia at angles (a) slightly less than the golden angle, (b) equal to the golden angle, (c) slightly more than the golden angle. Only in case (b) do they pack efficiently.

approach the next passage of the magical maze. What have we discovered so far?

Passage one draws upon two themes – one on the surface, one a little deeper. The surface theme is number theory, the strange properties of whole numbers. We saw them used in magical tricks, calendar games, and flower numerology.

The deeper theme is 'cycles' and 'near cycles'. The magic tricks, the calendar, and the swirling spirals of sunflower seeds all rely on the interplay between numbers and repetitive cycles. The most fascinating structures and ideas emerge when the cycles do not fit together perfectly – leap years, the drift of the seasons, the *bent* spokes of the seed-head, which make it possible to find two families of spirals and create a robust structure fit for a viable plant, rather than a flimsy set of spokes that would fall apart all too easily.

The magic of mathematics is subtle and deep. It reaches far beyond the simplest and most obvious of patterns. The mathematics of imperfections is just as useful – and in many ways *more* beautiful – than that of perfect simplicities.

JUNCTION TWO

On the wall of the second passage, just inside the gap where you turn right and go back on yourself, there is a mural.

It is a curious image.

A man is walking along the road with a gigantic bowl of porridge balanced on his head, and a broad smile on his rustic features. Ahead of him ambles his pet pig. Behind him, on a strong lead, prowls a huge black panther.

The mural has a title: FARMER WHATSIT TAKES HIS GOODS TO MARKET.

There is no signature, which is probably no surprise.

The story behind the mural is a familiar Mazeland folk-tale. Every school-child knows the long-running saga of Farmer Whatsit, his thousand and one trials and tribulations, and his eventual triumph.

In the relevant episode, the 97th in the saga, Whatsit is indeed taking his goods – panther, pig, and porridge – to market. He comes to the edge of a river, and stops. The ferryman has been called away unexpectedly, leaving a note pinned to the jetty: 'Please use small rowing-boat provided.' Whatsit looks around for the boat, and eventually spots it moored on the far side of the river. 'Oh, bother,' he says. (It is strongly believed by Mazeland scholars that this part of the story has been bowdlerised.) Obviously somebody has got there before him, and now the boat is on the wrong side. If he has to wait for someone to come the other way, he might be stuck for days.

The porridge will go cold. Nobody buys cold porridge at the Mazeland market. He might as well feed it to his pig, Swiney Todd.

Farmer Whatsit stamps his feet in annoyance. (Mazeland children all act out the stamping when they reach this part of the story.) Just before he completely loses his temper, he notices that the boat is tied to a long length of rope, and can be pulled back across the river. He does so, to find the oars resting inside it.

Whatsit smiles once more.

He steps tentatively into the boat, which wobbles in a most disturbing manner. It is very small, very flimsy, and very unstable. Whatsit knows a bit about boats, and he quickly realises that it will be dangerous to carry more than *one item of produce across at a time. The panther, the pig, or the porridge – but not two of them at once, let alone all three. The porridge can't row. Neither, for that manner, can Swiney Todd or the panther.*

   How can Farmer Whatsit and his goods cross the river?

Passage Two

PANTHERS DON'T LIKE PORRIDGE

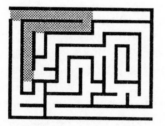

Elementary, my dear Whatsit – as another Mazeland story character, the infallible detective Shirley Combs, never actually said. Take the panther first, tie it up on the other side, come back to get the pig, and finally pick up the bowl of porridge.

Ah, yes, but ... Whatsit is about to load the panther when he sees Swiney Todd eyeing the porridge bowl and licking its chops in antici-pation. Swiney is a greedy beast, and Whatsit has a sudden vision of what will await him on his return.

An empty bowl, and a very fat, smug pig.

Then a worse vision assails him. If he leaves panther and pig alone together on the far bank – well, panthers are carnivores, and pigs are not renowned for their running skills. A fat, smug panther and a pathetic heap of bones, that's what will await his return from transporting the porridge.

There is just one bright spot: panthers don't like porridge. A good job, since porridge can run even less quickly than pigs – but the sight of a hungry panther prowling in search of a big steaming bowl of scrumptious porridge is not, on the whole, one you witness every day.

Whatsit seats himself beside the river, and begins to think the problem through.

It's not a new puzzle, of course. You've probably penetrated the disguise: it's more familiar as the wolf-goat-cabbage puzzle, with the wolf in the role of the panther, the goat playing Swiney Todd, and the cabbage filling in for the porridge bowl. I just thought it might be nice to change the characters a bit. The puzzle seems to have first been written down by the medieval mathematician Alcuin (735–804), although it probably went back even further in oral tradition☛. Similar puzzles can be found in many cultures.

CONCEPTUAL MAPS

I'm not really interested in this puzzle for its own sake, because it's not too hard to spot an answer, if you think everything through logically. What interests me is general methods for solving questions of this kind. One very effective approach is to represent all the possible actions as a maze, and try to find a route through it. It is a logical maze rather than a real one, touched with that magic genius of mathematical transformations in which a problem that seems unassailable in one form becomes trivial in another, logically equivalent one.

The idea is to represent the problem in a visual manner, using a diagram called a *graph*. A graph consists of a number of *nodes* (dots) linked by *edges* (lines), possibly with arrows on them. Each 'state' of the puzzle – position of the items of produce relative to the river – is represented by a node. Each 'legal' move between states is represented as an edge joining the corresponding nodes. If necessary, arrows can be added to the edges to show which is the starting state and which is the end state. The solution of the puzzle then reduces to tracing a path through its graph, starting from the initial state of the problem and finishing at the desired final state. The graph is a kind of conceptual map of the puzzle – a maze of possible states whose passages are the edges of the graph and whose junctions are its nodes.

That probably sounds a bit baffling, in the abstract, so we'll apply the method to the river-crossing problem and follow through stage by stage. But I thought you'd probably like to know where we're headed.

The first step is to simplify the problem, by reducing it to its essential features. Mathematics always works best when its raw materials are as simple as possible: extraneous information (such as 'how heavy was the pig?') should be resolutely excluded. The important thing here is which side of the river each of the three items is on. It's irrelevant where Whatsit is, or where the boat is, because those are free to move at will. The sole constraints are that the panther should not be left alone with the pig, and the pig should not be left alone with the porridge.

The essential features can be captured by introducing a nice, compact notation. Symbolism isn't everything in mathematics, but it often suggests useful ideas – and here the right symbolism pays dividends. I'll represent the position of a given item by a single digit, either 0 or 1. Here 0 represents *this* side of the river (where Whatsit is right now) and 1 represents the far side (where he wants to go).

If, for instance, all three items are on this side – as they are at the start – then we can list the states like this:

panther	0
pig	0
porridge	0.

If Whatsit then takes the pig across – which is a 'legal' move, since panthers don't like porridge – the pig is now on the far side, and the panther and porridge are on this side, so we can list the new states like this:

panther	0
pig	1
porridge	0.

These two states are connected by a legal move. Of course, what we want to end up with is

panther	1
pig	1
porridge	1,

but it's not yet clear how to achieve this.

Instead of writing the lists out in full, we can simplify the notation by just listing the three digits 0 or 1 in the order (panther, pig, porridge). So the three states just described are, respectively, $(0, 0, 0)$, $(0, 1, 0)$, and $(1, 1, 1)$.

How many possible states are there? Eight altogether:

$(0, 0, 0)$	all on this side
$(1, 0, 0)$	pig and porridge on this side, panther on far side
$(0, 1, 0)$	panther and porridge on this side, pig on far side
$(1, 1, 0)$	porridge on this side, panther and pig on far side
$(0, 0, 1)$	panther and pig on this side, porridge on far side
$(1, 0, 1)$	pig on this side, panther and porridge on far side

(0, 1, 1)   panther on this side, pig and porridge on far side

(1, 1, 1)   all on far side.

What are the possible legal moves? Well, starting from (0, 0, 0) we *could* move to (1, 0, 0), (0, 1, 0), or (0, 0, 1). Any move changes the state of just one item, so one entry in the list changes from 0 to 1 or vice versa. With three entries, there are always three moves to contemplate. However, in this case only one of them is 'legal' – that is, it avoids one item eating another while the farmer is busily rowing across the river. That's (0, 1, 0) – take the pig across and leave the panther to turn up its nose in disdain at the thought of eating a bowl of porridge. Let me list the possibilities, in Table 4.

Table 4

State   Can move to all these but only <u>underlined</u> moves are legal

(0, 0, 0)   (1, 0, 0)   <u>(0, 1, 0)</u>   (0, 0, 1)

That's one case; what about the rest? In general, the question to be answered when checking whether a move is legal is 'what does the farmer *leave behind*?' Bear in mind that *he* crosses the river along with whichever item has changed state – which is the panther in column 2, the pig in column 3, and the porridge in column 4. So the rules for legality are these:

- In column 2 the second and third entries must be different (pig and porridge are not left on the same side when panther is taken across).
- In column 3 anything is allowed (if pig is taken across it doesn't matter where panther and porridge are).
- In column 4 the first and second entries must be different (panther and pig not left on same side when porridge is taken across).

Now we can systematically complete the table for all the other seven states (Table 5).

Table 5

State   Can move to all these but only <u>underlined</u> moves are legal

(0, 0, 0)	(1, 0, 0)	<u>(0, 1, 0)</u>	(0, 0, 1)
(1, 0, 0)	(0, 0, 0)	<u>(1, 1, 0)</u>	<u>(1, 0, 1)</u>
(0, 1, 0)	<u>(1, 1, 0)</u>	<u>(0, 0, 0)</u>	<u>(0, 1, 1)</u>
(1, 1, 0)	<u>(0, 1, 0)</u>	<u>(1, 0, 0)</u>	(1, 1, 1)
(0, 0, 1)	<u>(1, 0, 1)</u>	<u>(0, 1, 1)</u>	(0, 0, 0)

$(1, 0, 1)$	$(\underline{0}, \underline{0}, \underline{1})$	$(\underline{1}, \underline{1}, \underline{1})$	$(\underline{1}, \underline{0}, \underline{0})$
$(0, 1, 1)$	$(1, 1, 1)$	$(\underline{0}, \underline{0}, \underline{1})$	$(\underline{0}, \underline{1}, \underline{0})$
$(1, 1, 1)$	$(0, 1, 1)$	$(\underline{1}, \underline{0}, \underline{1})$	$(1, 1, 0)$

So in fact, out of the 24 moves listed, precisely 8 are illegal and 16 legal. The 16 legal moves come in pairs: in each pair the same item is carried across the river in either direction. Thus we can move from $(0, 1, 0)$ to $(1, 1, 0)$ *or* from $(1, 1, 0)$ to $(0, 1, 0)$. In both cases the panther is taken across the river, but in the first case it goes from this side to the far side, and in the second case it comes back again.

Now we need to find a way to draw the graph of the puzzle – the diagrammatic representation of its states and legal moves. There will be eight nodes, corresponding to the eight states, and eight connecting lines, corresponding to the pairs of legal moves. The lines will not require arrows, since each is a two-way passage.

Figure 11 shows a neat way to do this. Legal moves are marked as solid lines, and – for information only – illegal ones are shown shaded. The graph has a striking geometric structure: a cube. In fact, we can think of the three numbers in the list of states as coordinates in three-dimensional space. There are three axes: a left/right one in the 'panther' direction, a front/back one in the 'pig' direction, and an up/down one in the 'porridge' direction. I know that porridge isn't usually a direction, but the figure shows what I mean. If we think of a list like $(0, 1, 1)$ as three coordinates in space, then it means 'go zero units along the panther direction, one unit along the pig direction, and one unit along the

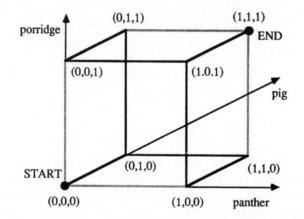

Figure 11   Picturing the river-crossing puzzle in pather–pig–porridge space.

porridge direction'. That is, don't move along the left/right axis, but go one unit to the back and one up. Now, the eight states are the vertices of a cube, the 24 legal or illegal moves are its twelve edges, traversed in either direction, and the 16 legal moves are what edges remain when we excise four of them (again traversed once in each direction).

More formally, the states of all three items are represented by a triple (P, p, π) in a conceptual three-dimensional *panther–pig–porridge space*. For example $(P, p, \pi) = (1, 0, 1)$ represents $P = 1$, $p = 0$, $\pi = 1$; that is, the panther on the far side, the pig on this side, and the porridge on the far side.

In this diagrammatic form, the puzzle becomes very simple. Can I start at vertex $(0, 0, 0)$ of the cube (all items on this side) and get to vertex $(1, 1, 1)$ of the cube (all items on the other side) by passing only along solid edges of the diagram? And of course the answer is 'yes'. Indeed, by distorting the diagram suitably without altering its connections, I can lay the edges out flat (Figure 12) and the solution stares us in the face. Two solutions, in fact – and only two if we avoid unnecessary repetitions.

Solution 1

$(0, 0, 0)$	start
$(0, 1, 0)$	take pig over
$(0, 1, 1)$	(return and) take porridge over
$(0, 0, 1)$	bring back pig
$(1, 0, 1)$	take panther over
$(1, 1, 1)$	(return and) take pig over.

Solution 2

$(0, 0, 0)$	Start
$(0, 1, 0)$	Take pig over
$(1, 1, 0)$	(return and) take panther over
$(1, 0, 0)$	bring back pig
$(1, 0, 1)$	take porridge over
$(1, 1, 1)$	(return and) take pig over.

THE LOGICAL MAZE

I've described this geometrical method at length because it applies to a broad range of puzzles, in which various objects must be rearranged while obeying certain rules, and the object is to get from some starting position to some finishing position.

Remember what we did. We wrote down *all* the states and legal moves

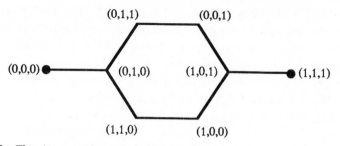

Figure 12   The river-crossing puzzle laid out flat.

(here it turned out to be helpful to have a systematic notation, but that's not essential). Then we formed a graph whose nodes correspond to states and whose edges correspond to legal moves. The solution of the puzzle is then a path through the graph that joins the start to the finish. Such a path is usually obvious to the eye, provided the puzzle is sufficiently simple for the entire graph to be drawn.

Puzzles of this type are really mazes, for a maze is just a graph drawn in a slightly different fashion. Metaphorically, they are logical mazes – you have to find the right sequence of moves to solve them. The graph turns the logical maze into a genuine maze, turning the metaphor into reality. The fact that solving the real maze also solves the logical maze is one of the magical features of the maze that is mathematics.

The graphical approach is applicable – in principle – to many puzzles, but often it is impractical to draw the graph in full. This happens in particular if the number of positions or moves is very large. For example, the graph of the 'Rubik's cube' puzzle has 43,252,003,274,489,856,000 nodes!

THE MINOTAUR'S ADVICE

It's not especially useful to reduce a puzzle to traversing a maze, unless you can solve the resulting maze. The traditional method used by mythic heroes is a ball of string. However, there are better methods …

'Next time I see Ariadne,' Theseus muttered darkly, 'I'll tell her she ought to have given me a longer ball of string! Only five junctions into the labyrinth, and it's run out.' Petulantly he flung the ball to the floor; then debated whether to go on, stringless, or go back and face the derision of those who waited outside. He swallowed manfully, unsheathed his sword, and moved on, all senses alert for the first sign of the terrible Minotaur.

There was a rather bovine sort of smell ... He followed it down the tunnels, deeper and deeper into the labyrinth. The tunnels all looked so *similar* ...

'Lost?'

Through the gloom he made out a dim shape at his shoulder. Another would-be hero, no doubt. 'Yes,' he said hopelessly.

'Me too,' said the voice sympathetically. 'Been stuck in here for centuries.'

'You hunting the Minotaur too?' asked Theseus.

'Good grief, man, I *am* the Minotaur! Oh, put that silly pigsticker away! If I wanted to eat you I'd have done so by now.'

The light – which pervaded the labyrinth from unseen sources – improved a little, and Theseus could see his companion's face. More like a Guernsey cow than a bloodthirsty bull, though it did have the obligatory horns.

'Are you any good at getting out of mazes?' asked the Minotaur hopefully.

'Good at getting *into* them,' said Theseus. 'I guess the trick is to reverse the process.'

'I've tried every method I've ever heard of,' said the Minotaur. 'One way to solve a maze is to look at a map and block off all the dead-ends.'

'Do you have a map?' asked Theseus, perking up.

'Of course not,' said the Minotaur indignantly. 'You think when Daedalus built the place he lodged a copy of the architect's drawings with the Cretan County Council? Pah! If I had a map, I'd have been out of here long ago.'

'Fat lot of good blocking off dead-ends is, then,' observed Theseus.

'Well, you could try the old left-hoof-on-wall trick. When you first enter the labyrinth – or any maze, come to that – you put your left hoof on the wall, and keep it there. That guarantees that eventually you'll find your way out again.'

'I don't have a hoof.'

'It's too late now, anyway,' said the Minotaur sadly. 'It's only guaranteed to work if you use it from the very beginning. You didn't, by any chance ... no, of course not.'

'Do you know anything else about mazes?'

The Minotaur scratched a horn with one hoof. 'Well ... the most important thing about a maze is how the junctions connect up. The lengths of the tunnels don't matter—'

'I'm not sure I can agree with that,' said Theseus ruefully, rubbing at his corns.

'Well, they don't affect the *path* you take to get out, just its length, OK? Now, we can represent the topological essentials by a graph: its nodes correspond to the junctions in the maze, its edges correspond to the tunnels. Then the problem of getting out of a maze – or of finding a particular place within it – becomes that of traversing a graph from one node to another.'

'I think we know all that,' said Theseus.

'Yes, but, there's a really neat theorem that tells you when it's possible.'

'Terrific!' said Theseus. 'I adore theorems. Pythagoras told me a real stunner once, something about a square hippopota—'

'Not now, Theseus. This is the theorem about graphs. Two nodes can be joined by a continuous path if and only if they lie in the same connected component of the graph.'

'Hmmph,' said Theseus doubtfully. 'What's a connected component?'

'The set of all nodes that can be reached from a given one by a continuous path,' said the Minotaur proudly.

'Ah. Let me see if I've got this right. What you're saying is that two nodes can be joined by a continuous path if and only if there exists a continuous path that joins them?'

The Minotaur was deeply offended. 'Well … you could put it like that, but it seems to me you're trivialising an important concept.'

'That may be so. But I think we need something a little more constructive.'

'Oh, right. You want a maze-threading algorithm.'

'Do you mean the fabulous *Algorithmos labyrinthoi*, the ninety-headed monster with razor teeth and a gorgon haircut that stalks the ancient passages?'

'No, Theseus, it's not a mythical beast. It's just a word that people are going to invent. It's named after Muhammad ibn Musa abu Abdallah al-Khorezmi al-Madjusi al-Qutrubilli. "Al-Khorezmi" became "al-Gorizmi", then "algorism", and finally "algorithm". It's used to describe a specific procedure to solve a problem.'

DEPTH FIRST SEARCH

A very efficient maze-threading algorithm was invented around 1892 by M. Trémaux, and rediscovered nearly a century later by J. Hopcroft and

R. Tarjan. It is called the Depth First Search algorithm. The algorithm visits all nodes in the same connected component as the starting node; in particular, it can be terminated when it hits any selected 'exit' node.

Here are the instructions in the algorithm:

- Begin at any chosen node.
- Visit any adjacent node that has not yet been visited.
- Repeat this as far as possible.
- If all adjacent nodes have been visited already, backtrack through the sequence of nodes that have been visited until you find one that is adjacent to an unvisited node, then visit that one.
- Delete any edge that has been backtracked.
- Repeat until you return to the starting node and there are no unvisited nodes adjacent to it.
- Then you have visited all nodes in the connected component of the graph that contains the starting node.

Depth First Search is especially appropriate for threading mazes, because it is possible to use it without having a map of the maze. It involves only local rules at nodes, plus a record of nodes and edges already used, so you can explore the graph and traverse it as you go. The name indicates the basic idea: give top priority to pushing deeper into the maze. The number of steps required is at most twice the number of passages in the maze.

As we've seen, many puzzles (panther–pig–porridge being one) involve the movement of people, animals, or objects from a given position to a selected finishing position, subject to various rules. So do a lot of practical problems, for example scheduling the movement of the aircraft operated by an airline. Most practical problems involve additional considerations – such as the expense of a given 'move' – but their solutions can be obtained by very similar methods to those I'm applying to the puzzles. Such puzzles can be reduced to traversing mazes, and provided that the number of possible positions is finite, Depth First Search can be employed.

It then takes the following form:

- Whenever you first come to a new position (including the initial position at the start of the puzzle), list all possible positions that can be reached from it.
- From the starting position, make any move that leads to a 'new' position (one that has not yet occurred) at random.
- Repeat this as far as possible.

- If all possible moves lead to 'old' positions, backtrack through the sequence of moves that have been made until you find a position from which it is possible to move to a 'new' position. Then make that move.
- Henceforth ignore any move that has been backtracked.
- Repeat until you either reach the desired position, or return to the starting position with no moves available, in which case the puzzle is impossible.

To show you how this method works, I'm going to use Depth First Search to solve the panther–pig–porridge puzzle. This time I'll represent the participants by F = Farmer Whatsit, P = panther, p = pig, π = porridge. Positions are represented by symbols such as [Fp ‖ Pp], where ‖ represents the river – so here farmer and pig are on the left bank, panther and porridge on the right. How Depth First Search works out is shown in Table 6.

Table 6

step	current position	possible moves	move made	comments
1	[FPpπ ‖ —]	[Pπ ‖ Fp]	[Pπ ‖ Fp]	Starting position
2	[Pπ ‖ Fp]	[FPpπ ‖ —]		Old position
		[FPπ ‖ p]	[FPπ ‖ p]	Forced by algorithm
3	[FPπ ‖ p]	[Pπ ‖ Fp]		Old position
		[P ‖ Fpπ]	[P ‖ Fpπ]	Random choice of available moves
		[π ‖ FPp]		Other possible choice
4	[P ‖ Fpπ]	[FPπ ‖ p]		Old position
		[FPp ‖ π]	[FPp ‖ π]	Forced by algorithm
5	[FPp ‖ π]	[P ‖ Fpπ]		Old position
		[p ‖ FPπ]	[p ‖ FPπ]	Forced by algorithm
6	[p ‖ FPπ]	[FPp ‖ π]		Old position
		[Fπp ‖ P]	[Fπp ‖ P]	Random choice (bad move, but it shows
		[Fp ‖ Pπ]		how the algorithm copes)
7	[Fπp ‖ P]	[p ‖ FPπ]		Old position
		[π ‖ FPp]	[π ‖ FPp]	Forced by algorithm
8	[π ‖ FPp]	[FPπ ‖ p]		Old position ⎫ backtrack required!
		[Fπ ‖ P]		Old position ⎭
9	[Fπp ‖ P]	*see* step 7		Backtrack to step 7
10	[p ‖ FPπ]	*see* step 6		Backtrack to step 6
			[Fp ‖ Pπ]	Alternative move from position 6
11	[Fp ‖ Pπ]	[p ‖ FPπ]		Old position
		[— ‖ FPpπ]	[— ‖ FPpπ]	Forced by algorithm
12	[— ‖ FPpπ]			Finish

The path that the algorithm tracks through the graph is shown in Figure 13, but the algorithm does not require this graph to be drawn in advance: instead, sections are explored as needed. The solution takes a 'wrong turn' at step 6 but the algorithm successfully corrects this by backtracking.

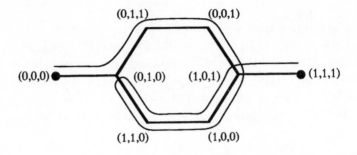

Figure 13   How Depth First Search tracks its way through Figure 12.

To make sure you've grasped the principle, you might try to solve the pouring problem of Figure 14 using Depth First Search. For the answer, see the 'Pointers' section☞.

NOT A LOT OF PEOPLE KNOW THIS, BUT ...

'Brilliant!' yelled Theseus. 'This idea will secure our freedom! Yes, Minotaur, I have not forgotten your pessimistic theory that we are trapped for ever by design of the gods – but not even Olympus can interfere with the workings of logic, or the gods would destroy the very basis of the universe, and with it their own purpose. We are saved! All we have to do is carry out a Depth First Search of the labyrinth!'

But the Minotaur seemed unenthusiastic. 'Trouble is, Theseus, to apply Depth First Search you need some way to mark each junction so that you know you've been there before, and where you came from. Did Ariadne happen to include a stick of chalk along with the string?'

Theseus admitted that chalk was unaccountably omitted from his equipment. 'I've got a sword, though. I'll scratch the walls.'

'No chance, these rocks are as hard as diamond.'

'Drat.'

They sat in silence, communing with their own dark thoughts. 'I guess you've invented the MUPS algorithm, the Minotaur's Universal Puzzle

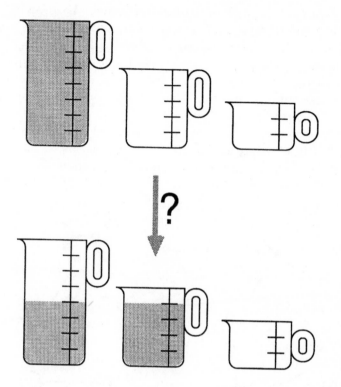

Figure 14   Split the eight litres of water into two equal parts, as indicated, by using Depth First Search.

Solver,' said Theseus. 'But what we really need is a universal method that gets us out of any maze whatsoever, and doesn't involve making any marks on the walls of the labyrinth.'

'Not a hope,' said the Minotaur. 'Nobody would be able to pronounce the MTGUOOAMWADIMAMOTWOTL algorithm.'

'Sarcastic old cow, aren't you,' said Theseus bitterly. Then he leaped to his feet in excitement. 'Wait! Your long sequence of letters has given me an idea … yes! Look, we know that there exists a sequence of left and right turns that will get you out of any particular maze. So maybe there's a sequence of left and right turns that gets you out of any maze whatsoever. Just one sequence – the same for all mazes! Something like LLLRRLRLRRLRRRRRL … It will have to be infinitely long because there are infinitely many possible mazes. I'll call it the TUMS algorithm – Theseus's Universal Maze Solver.'

'Nonsense! For a start, some junctions might have more than two possible paths. If you need to go straight on, then a choice of only left or right won't help, will it? And what happens at dead-ends?'

'I can get round that easily,' said Theseus, grinning from ear to ear. 'The choice will be between just "right" and "left" provided that exactly three tunnels meet at each junction. The one you approach the junction by, plus two others that fork.'

'But there may not always be three,' protested the Minotaur.

'Ah, but I can reduce any graph, and hence any maze, to one that has only triple junctions. I just replace a multiple junction by a ring of linked

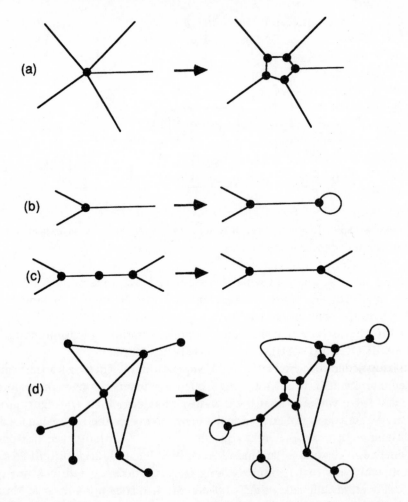

Figure 15   How Theseus reduces any maze to one with only triple junctions.

triple junctions (Figure 15(a)), and add a closed loop on to every dead-end (Figure 15 (b)). Oh, and if there's a "fake" junction at which only two edges meet, I delete it (Figure 15(c)). Now, from any graph I can produce a modified graph with only triple junctions (Figure 15(d)). If I can traverse the modified graph, then I can easily reconstruct a way to traverse the original graph by contracting my additions down to points.'

That got rid of the Minotaur's objection, but it still left one minor difficulty – *finding* a universal maze-solving sequence of left and right turns.

'Hmmph,' said Theseus eventually. 'Suppose I make a list of all possible mazes—'

'Yes! Yes!' interrupted the Minotaur, eyes sparkling with the splendour of his sudden insight. 'First you list all mazes – or graphs – with two nodes, then those with three, then four, and so on, counting only the graphs that have triple junctions, of course. Then you string all the lists together, one after the other.'

Theseus wondered whether that was possible. 'What if there are infinitely many graphs for some number of nodes? Then your list goes off to infinity and there's nowhere "after" it to stick the next list on.'

'Ah, but the number of distinct graphs with a given finite number of vertices is always finite.'

'Why?'

'You can represent any graph on n nodes by an $n \times n$ array of 0's and 1's. You just number the nodes from 1 to n, and then make the entry in row r and column c of the array equal to 1 if nodes r and c are joined by an edge, and 0 if not. The array completely specifies the connections of the graph. And the number of possible $n \times n$ array of 0's and 1's is 2^{n^2}, which is finite.'

'Right … So the number of different graphs on n nodes is less than or equal to that, hence also finite,' said Theseus. 'We're on to something, Minotaur – I see it all now. When you've listed all the possible maze-graphs in order of complexity, you look at the first one, choose a starting place and an exit point in it, and work out what sequence of lefts and rights will get you out. That will be a finite sequence.'

'Yes, but what if you start somewhere else in the maze? Or need to find a different exit?'

'That's the clever bit. You choose a different start and exit in the first graph, and see where the previous sequence would have got you to if you hadn't been in the place you'd originally started from …'

There was a long pause while the Minotaur digested this. Theseus was about to speak when, surprisingly, the Minotaur interrupted.

'If that gets you out, fine; if not, you add on another sequence that will. Of course. And then you go on to do it for all the other combinations of starting points and exits in the first maze?'

'Yes. Oh, there's probably a more efficient way, but it works. And when you've finished that, you go on to the second maze, and extend the sequence to deal with all possible starting points and exits, one after the other; then the third maze, and so on.'

The Minotaur thought hard before asking slowly: 'There are infinitely many positions and mazes, aren't there, Theseus?'

'Yes, but as you yourself observed, we can still write them in order as an infinite list, and that's all we need.'

'Yes, Theseus, I see that. But if I can list this infinite set of mazes and starting positions, then I can list it in *any* order, can't I?'

'I guess ...'

'Well, then, can't I list them so that all the mazes that you can get out of by making only left turns come first? So my sequence begins with infinitely many left turns, LLLLLL ...?'

Theseus was vaguely worried (and so should you be – why?☞), but before he could say so, the Minotaur had rushed off into the maze. His voice, fainter and fainter, echoed off the tunnel walls: 'Left ... left ... left ... left ...'

In a sudden panic, no longer worried about the logicality of the argument but scared that he might be left behind, Theseus also rushed off, but instead he could be heard yelling: 'Right ... right ... right ... right ...'

After an hour or two, Theseus stopped turning right. Looking carefully around to make sure he was not observed, he gingerly placed his left hand on the wall and set off through the labyrinth. After a while he passed the Minotaur, limping along with both left hooves on the opposite wall, travelling in the opposite direction. Both looked highly embarrassed and avoided eye contact as they passed, pretending to be deep in thought.

TOWER OF HANOI

My all-time favourite graph comes from another traditional puzzle, the Tower of Hanoi. This puzzle was marketed in 1883 by the French mathematician Édouard Lucas, under the pseudonym M. Claus. In *La Nature* in 1884, M. De Parville described it in romantic terms:

In the great temple at Benares, beneath the dome which marks the centre of the world, rests a brass plate in which are fixed three diamond needles, each a cubit high and as thick as the body of a bee. On one of these needles, at the creation, God placed sixty-four discs of pure gold, the largest disc resting on the brass plate, and the others getting smaller and smaller up to the top one. This is the Tower of Bramah. Day and night unceasingly the priests transfer the discs from one diamond needle to another according to the fixed and immutable laws of Bramah, which require that the priest on duty must not move more than one disc at a time and that he must place this disc on a needle so that there is no smaller disc below it. When the sixty-four discs shall have been thus transferred from the needle on which at the creation God placed them to one of the other needles, tower, temple, and Brahmins alike will crumble into dust, and with a thunderclap the world will vanish.

The Tower of Hanoi puzzle is just like the Tower of Brahmah, but played with fewer discs – usually eight or so. The puzzle is truly a maze, and our first objective is to map out that maze and see what it looks like. To do so, it helps to think about even smaller versions of the puzzle, with only two or three discs. They're no fun to play, but they yield a lot of insight. Again, our aim is not so much to hack our way to a solution, as to argue our way towards a general method. Mathematicians have a catch phrase for this trick: 'let $2 = n$'. Anyone can specialise from the general to the particular ('let $n = 2$'); the true art of the mathematician is to go the other way, and generalise an idea from a particular case to all of them. By learning to navigate their way through small but representative parts of the magical maze, mathematicians often find out how to fight their way through the whole thing.

For definiteness, consider 3-disc Hanoi, that is, the Tower of Hanoi with just three discs (Figure 16). I'm going to construct the graph of the puzzle, just as I did for panther–pig–porridge. As before, we must first find a way

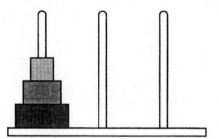

Figure 16   Tower of Hanoi with three discs.

to represent all possible positions, and then work out all the legal moves between them. Finally, we draw the graph.

How can we represent a position? Anything simple and systematic will do. One way is to number the three discs 1, 2, 3, with 1 the smallest and 3 the largest. Number the needles, too: 1, 2, 3, from left to right. A typical state of the puzzle has, say, disc 1 on needle 2, disc 2 on needle 1, and disc 3 on needle 2. Once we know which discs are on which needles, we have completely determined the position, because the rules imply that bigger discs, such as 3, must be underneath smaller discs, such as 1. So we can encode the state using the sequence 212, where the three digits in turn represent the needles for discs 1, 2, and 3. It follows that each position in 3-disc Hanoi corresponds to a sequence of three digits, each being 1, 2, or 3. To make this clear, Figure 17 gives several examples.

All codes can occur, so, in particular, there are precisely $3 \times 3 \times 3 = 27$ different positions in 3-disc Hanoi. But what are the permitted moves?

The smallest disc on a given needle must be at the top. It thus corresponds to the *first* appearance of the number of that needle in the sequence. If we move that disc, we must move it to the top of the pile on some other needle – that is, we must change the number so that it becomes the first appearance of some other number.

For example, in the position 212 above, suppose we wish to move disc 1.

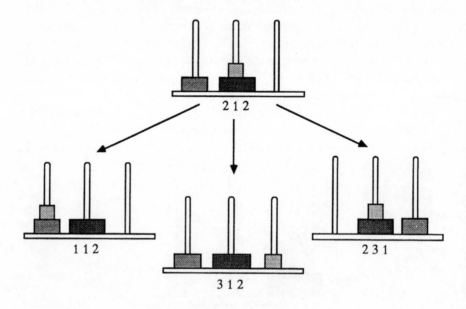

Figure 17   Typical moves in 3-disc Hanoi.

This is on needle 2, and corresponds to the first occurrence of 2 in the sequence. Suppose we change this first 2 to 1. Then this is (trivially!) the first occurrence of the digit 1; so the move from 212 to 112 is legal. So is 212 to 312, because now the first occurrence of 3 is in the first place in the sequence.

We may also move disc 2, because the first occurrence of the symbol 1 is in the second place in the sequence. But we cannot change it to 2, because 2 already appears earlier, in the first place. A change to 3 is, however, legal. So we may change 212 to 232 (but not to 222).

Finally, disc 3 cannot be moved, because the third digit in the sequence is a 2, and this is not the first occurrence of a 2.

To sum up: from position 212 we can make legal moves to 112, 312, and 232, and only these. Try it with actual discs.

We can list all 27 positions and all possible moves by following the above rules. The result is shown in Table 7.

Table 7

start here	finish on any of these		
111	211	311	
112	212	312	113
113	213	313	112
121	221	321	131
122	222	322	132
123	223	323	133
131	231	331	121
132	232	332	122
133	233	333	123
211	111	311	231
212	112	312	232
213	113	313	233
221	121	321	223
222	122	322	
223	123	323	221
231	131	331	211
232	132	232	212
233	133	333	213
311	111	211	321
312	112	212	322
313	113	213	323
321	121	221	311

322	122	222	312
323	123	223	313
331	131	231	332
332	132	232	331
333	133	233	

The table has a strange feature: all but three positions lead to exactly three legal moves, but the other three positions lead to only two legal moves. The wise mathematician always worries about strange features, if only because they might indicate a mistake. Here, however, all is well: these three exceptional positions really are special. They are the positions in which all the discs are on one needle, the other two being empty. From such positions, only the top disc can be moved, and there are only two places to which it can go.

The next task requires care and accuracy, but little thought. Draw 27 dots on a piece of paper, label them with the 27 positions, and draw lines to represent the legal moves. My first attempt at this ground to a halt in a mess of spaghetti. A bit of thought, rearranging the nodes and edges to avoid overlaps, led to Figure 18. Something that pretty can't be coincidence! The mathematician's antennae begin to twitch. The pattern has suddenly become more significant than the puzzle that gave rise to it. We have found an unexpected feature in the logical maze of 3-disc Hanoi. If

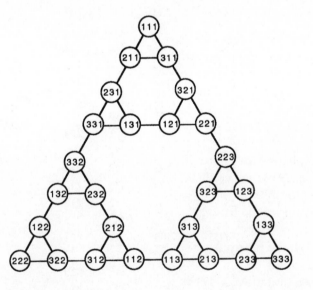

Figure 18   The graph of 3-disc Hanoi.

we can understand that feature, it may give us power over the general problem.

But before we investigate why the graph has such a regular form, let's observe that it provides an easy answer to 3-disc Hanoi. To transfer all three discs from needle 1 (position 111) to needle 2 (position 222), we merely run down the left-hand edge of the graph, making the moves

$$111 \rightarrow 211 \rightarrow 231 \rightarrow 331 \rightarrow 332 \rightarrow 132 \rightarrow 122 \rightarrow 222.$$

By consulting the graph, it is clear that we can get from any position to any other – a fact that wasn't at all obvious when we started – and in most cases it is also pretty clear what the quickest route should be.

A MILITARY SOLUTION

On to a deeper question: what is the explanation for the remarkable structure of Figure 18?

The graph consists of three copies of a smaller graph, linked by three single edges to form a triangle. Each smaller graph in turn has a similar triple structure. Why does everything appear in threes, and why are the pieces linked in this manner?

If you work out the graph for 2-disc Hanoi, you will find that it looks exactly like the top third of Figure 18. Even the labels on the vertices are the same, except that the final 1 is deleted. In fact it is easy to see why, without working out the graph again. You can play 2-disc Hanoi with three discs: just ignore disc 3. Suppose disc 3 stays on needle 1. Then we are playing 3-disc Hanoi, but restricting attention to those 3-digit sequences that end in 1, such as 131 or 221. But these are precisely the sequences in the top third of the figure. Similarly, 3-disc Hanoi with disc 3 fixed on needle 2 – that is, disguised 2-disc Hanoi – corresponds to the lower left third, and 3-disc Hanoi with disc 3 fixed on needle 3 corresponds to the lower right third.

This explains why we see three copies of the 2-disc Hanoi graph in the 3-disc graph. The figure shows that these three subgraphs are joined by three single edges in the full puzzle; and a little further thought shows why. To join up the subgraphs, we must move disc 3. When can we do this? Only when one needle is empty, one contains disc 3, and the other contains all the remaining discs! Then we can move disc 3 to the empty needle, creating an empty needle where it came from, and leaving the other discs untouched. There are six such positions, and the possible moves join them in pairs.

The same argument works for any number of discs. The graph for 4-disc Hanoi, for example, consists of three copies of the 3-disc graph, linked at the corners like a triangle. Each subgraph describes 4-disc Hanoi with disc 4 fixed on one of the three needles; but such a game is just 3-disc Hanoi in disguise. And so on. We say that the Tower of Hanoi puzzle has a *recursive* structure. By this, we mean that the solution to $(n+1)$-disc Hanoi can be determined from that for n-disc Hanoi according to a fixed rule. The recursive structure explains why the graph for $(n+1)$-disc Hanoi can be built from that for n-disc Hanoi. The triangular symmetry arises because the rules treat needles 1, 2, and 3 in exactly the same way. You can deduce the graph for 64-disc Bramah, or any other number of discs, by repeatedly applying this rule to the graph for 0-disc Hanoi, which is a single dot! For example, Figure 19 shows the graph for 5-disc Hanoi, drawn by applying this recursive structure. Brains instead of brawn – it would take hours to work it out by listing all 243 possible positions and finding all the moves between them ... and you'd probably make several mistakes along the way.

There's a nice way to exploit the recursive structure of the Tower of Hanoi to solve the puzzle completely. You may recall the story of an army major being trained to put up a flagpole. The recommended method is this: find a sergeant, and give him the order 'Sergeant, put up this

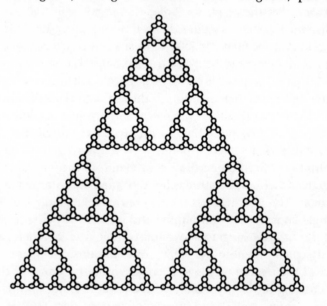

Figure 19   The graph of 5-disc Hanoi.

flagpole.' The recursive solution to the Tower of Hanoi works in just this fashion, so I call it the *army method*. For definiteness, assume that the discs start on needle 1 and must end on needle 2. The army has a large number of privates, who have been trained to solve 1-disc Hanoi but nothing more ambitious:

'Bloggs. Move the disc to needle 2 while keeping it in order of size – yes Bloggs, I *know* you can't change the order of one disc, but I wouldn't be surprised if you 'orrible shower found a way to muck even *that* up.'

One private doesn't even know this, but he does know how to move the largest disc to a vacant needle. His name is Private Aggs. With the aid of Bloggs and Aggs, the corporals can solve 2-disc Hanoi:

'Private Bloggs. Move the smaller disc to needle 3 as in 1-disc Hanoi. Private Aggs. Move the larger disc to needle 2. Private Bloggs. Move the smaller disc to needle 3 as in 1-disc Hanoi.'

The sergeants can solve 3-disc Hanoi:

'Corporal Cloggs. Move the top two discs to needle 3 as in 1-disc Hanoi. Private Aggs. Move the larger disc to needle 2. Corporal Cloggs. Move the top two disc to needle 3 as in 1-disc Hanoi.'

The lieutenants can solve 4-disc Hanoi:

'Sergeant Doggs. Move the top three discs to needle 3 as in 1-disc Hanoi. Private Aggs. Move the larger disc to needle 2. Sergeant Doggs. Move the top three discs to needle 3 as in 1-disc Hanoi.'

You should now be able to work out how Captain Eggs, Major Foggs, Colonel Goggs, and n-star Generals Hoggs and so forth solve their own levels of the puzzle. To solve n-disc Hanoi for $n \geqslant 8$, you need an $(n - 7)$-star general.

I once tried to convince Yorkshire Television to enact this method for 5-disc Hanoi, using real soldiers. It would have been delightfully chaotic. Unfortunately, YTV decided it would put undue strain on the audience's attention span.

You can use the graph to answer all sorts of questions about the puzzle. For example, it follows recursively that the graph is connected – all in one piece. Why? Well, the graph for 3-disc Hanoi is connected. You get the graph for 4-disc Hanoi by *joining* three copies of the graph for 3-disc Hanoi together. So that's all in one piece too. The same goes for the graphs for 5-disc Hanoi, 6-disc Hanoi, and so on. The connectedness of the graphs means that you can always move from any position to any other.

You can also use the recursive structure to show that, no matter how many discs there are, the shortest path from the standard starting position (all on one peg) to the usual finishing position (all on another peg) runs straight along one edge of the graph, so it takes $2^n - 1$ moves to solve the n-disc puzzle, and no fewer.

THE MATHEMATICAL MAZE

The analogy with threading a maze runs deeper than games and puzzles. It illuminates the whole of mathematics. Indeed, one way to think about mathematics is as an exercise in threading an elaborate, infinitely large maze. A logical maze. A maze of ideas, whose pathways represent 'lines of thought' from one idea to another. A maze which, despite its apparent complexity, has a definite 'geography', to which mathematicians are unusually attuned.

The image of mathematics as a logical maze began to take form around the end of the nineteenth century, when a number of leading mathematicians started to worry about the logical foundations of their subject. They had good reason to be worried, too. As the conceptual structures of interest to mathematicians became more and more elaborate – for reasons of efficiency, generality, and deductive power – the danger of pushing the subject too far and too fast became apparent. In particular the area known as analysis – the intellectual descendant of calculus, hovering uneasily around the ideas of the infinitely large and the infinitely small – over-extended itself during the early nineteenth century, and spent several decades in a state of crisis. Different mathematicians managed to 'prove' apparently valid theorems that contradicted each other; cherished assumptions turned out to be false. It was a time of great confusion, and it arose because for a period everybody was so busy moving their subject forward that they got sloppy.

Today, many people – physicists and engineers are among the worst – often complain about mathematicians' excessive (they say) emphasis on proofs. If they knew the history of mathematics better, they would stop complaining, and begin to understand that without proof – very stringent, logically impeccable proof – mathematics becomes so unreliable that it is worse than useless.

At any rate, a point was reached when several top mathematicians decided that it was high time they set their subject to rights, rebuilding everything from the ground floor up and placing the entire edifice on firm foundations. Following a lead set by Euclid some two and a half millennia earlier, they founded mathematics in a small collection of basic assumptions – *axioms*.

In everyday language, an axiom is often held to be a 'self-evident truth'. That phrase betrays some awfully sloppy thinking. A truth can be evident, but *self*-evident? Evidence is what convinces *people* that something is true. So what on earth is 'self-evident' supposed to mean? A truth that convinces *itself*?

I guess that what's intended by the phrase is a statement so obviously true that absolutely nobody would be able to imagine contesting it – one that provides its own evidence, merely by being stated. But I don't believe that such statements exist. Indeed, many of the most important steps in the development of mathematics required people to possess enough imagination to challenge universally accepted 'self-evident' truths, and to show that they were actually nonsense. Examples include the beliefs that parallel lines necessarily exist, that negative numbers cannot have square roots, and that a times b is always equal to b times a, no matter what manner of beast a and b might be.

To those who were seeking foundations for mathematics, axioms were much more prosaic. They weren't truths at all, let alone evident ones, and certainly not *self*-evident ones – assuming that such a slogan means any-thing at all. Axioms were *assumptions*. A place to start. A collection of statements that mathematicians agreed to accept. You are free to challenge them if you wish, but even if you do, you won't change mathematics (though you might create some more, branching off in a new direction). From the axiomatic viewpoint, mathematics consists of the *deductions* that are made, once the axioms are accepted as a starting point. It's like a game. If you want to play football, then you follow the rules of football. Of course you are free to change the rules – but then you're playing a different game.

The axiomatic approach to mathematics goes back at least to the ancient Greeks, but seems not to go back further. Its advantages are many. One is economy of thought; another is elegance. A third is that, provided your chain of logic is correct, any flaws revealed as a result of lengthy deductions must be traceable back to the axioms. Suppose, for instance, that you set up axioms for arithmetic, and five thousand pages later you prove that $2 + 2 = 5$. Suppose, further, that you make no mistakes in logic along the way. Then the axioms themselves must be faulty. This, perhaps, is the sense in which axioms are 'foundations'. If a building is erected in a flawless manner and subsequently falls down, then there must have been something wrong with its foundations. I don't suggest you take this analogy *too* literally – what about earthquakes, for example? But with a sympathetic interpretation, it captures the essence of the axiomatic method.

The favoured axiom systems for mathematics are very, *very* rudimentary. Euclid's concepts of a line or a point are extremely esoteric compared to the current axioms. Even concepts such as '1' or '2' are esoteric. The axioms concentrate on basic steps in logical deductions, and on simple properties of collections of things – *sets*. Once you have sets, you can build your way up to complicated ideas like '2', or 'line', or 'differential equation'. The foundations of mathematics, like those of Canterbury Cathedral, are a long way below ground, invisible to the casual observer, and of no immediate interest to those who are working on the rest of the building. Take the foundations away, however, and the interest suddenly becomes keen. This is what foundations are for.

From the axiomatic viewpoint, mathematics consists of logical statements – all of which can be reduced, eventually, to statements about sets. Because numbers are more familiar to most of us, I shall use arithmetical statements as examples. The statements are the nodes of a maze – the maze of mathematical ideas. The pathways of the maze are elementary logical deductive steps, which lead from one statement to another. For example, the statement '2 + 2 = 4' leads to such statements as '2 + 2 + 1 ≡ 4 + 1' or '(2 + 2) × 5 = 4 × 5'. Figure 20 shows a tiny, tiny piece of the mathematical maze.

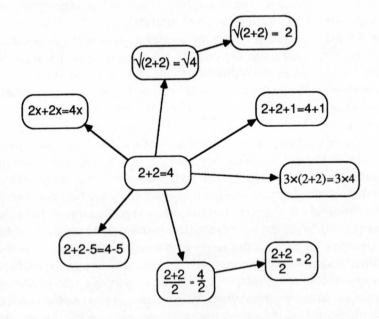

Figure 20   A tiny piece of the mathematical maze.

What are the axioms? They are the *entrance* to the maze – the place(s) you can start from.

What is the goal of the maze – the centre or the exit? If you are a research mathematician, the goal is the theorem that you are trying to prove. The maze goes on for ever, there is no ultimate exit; but there are all sorts of delightful little bowers and summerhouses and birdcages to be found within the maze, if you have the knack of following suitable paths.

HAVE I GOT HYPOTENUSE FOR YOU

There's exactly one theorem that everybody remembers. They may not remember what it says, but they remember its name: Pythagoras's theorem. It's just one of those things that stick in the mind. I have no idea why, but 'Pythagoras' is a curious name, and it's the first theorem most of us meet that's named after a person. It tells us that *people prove new theorems.*

Or, at least, that they did, two and a half thousand years ago. They still do, in fact, by the bucketful, but we don't often get reminded of that.

We don't know for sure that Pythagoras existed as a historical figure, and even if he did we don't know that he was the first to prove the theorem that bears his name. No matter: let's pretend he did exist, and that he proved his theorem. OK, then: imagine Pythagoras seeking a proof that the square on the hypotenuse of a right triangle is equal to the sum of the squares on the other two sides. (Yes, you remember now. But what's a hypotenuse? Never mind.) Pythagoras has chosen to wander through that part of the mathematical maze that we know as 'geometry'. From the modern point of view he is already far, far away from the entrance, the axioms; but we know, thanks to the efforts of people such as David Hilbert and Bertrand Russell, that there are lengthy logical pathways connecting today's axiomatic entrance to the statements upon which Euclid founded classical geometry. To Pythagoras, Euclid's statements *were* the entrance; but from that stage onwards he and the moderns are threading the same maze. Pythagoras's problem is to find a chain of logical deductions that leads from what has already been proved to what he *hopes* can be proved – his famous observation about squares and hypotenuses.

The particular path that Euclid consigned to posterity is a rather technical sequence of statements about triangles, and it wouldn't help much to show it to you. Instead, let me show you my favourite proof, one that uses so few raw materials that it will convince almost anybody.

Let's recall the statement of the theorem. It is about a right triangle – that just means a triangle having one angle equal to 90º. The side of the triangle opposite the right angle is called the *hypotenuse*. This is based on the Greek for 'stretching under', which the hypotenuse does if you put the right angle at the top of the triangle. It is jargon, but jargon that is useful to geometers.

Pythagoras, for reasons that can never be known but may well have had something to do with tiling patterns, was led to consider the squares whose sides are the three sides of the triangle. And he became convinced that the area of the largest of these three squares is *exactly* equal to the combined areas of the other two. Euclid demonstrates Pythagoras's contention using a figure that generations of European schoolchildren called 'Pythagoras's pants', because it looks like a garment hung on a washing-line. I want you to think about a different picture: Figure 21(a). Unlike Pythagoras's pants, it involves four copies of the same right triangle, arranged to form a big square. That's not one of the squares that interested Pythagoras, but it plays a key role none the less. One of the three squares that interested Pythagoras appears in the centre of the figure, tilted and shaded. It is the star of the theorem, 'the square on the hypotenuse'.

But the other two squares that interested Pythagoras don't appear in this figure. To bring them into play, we replace the figure with another one, Figure 21(b). Then the other two squares are the little one at top left and the medium-sized one at bottom right – again shown shaded.

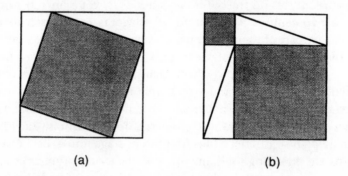

(a)   (b)

Figure 21   Proving Pythagoras: (a) the square on the hypotenuse; (b) the squares on the other two sides.

So?

Both pictures have features in common:

- They each form a big square. (In fact, the big square is the same size in both pictures: its side is the sum of the two sides of the right triangle that are *not* the hypotenuse. This is pretty obvious from the first figure, and only a little less so from the second.
- They each contain four copies of the right triangle.

However, there are differences:

- The first picture contains four copies of the right triangle *plus* the square on the hypotenuse.
- The second picture contains four copies of the right triangle *plus* the squares on the other two sides (not the hypotenuse).

Hmmm.

Since both pictures are, overall, the same size, it must follow that four copies of the right triangle *plus* the square on the hypotenuse have the same total area as four copies of the right triangle *plus* the squares on the other two sides.

If we deduct the four copies of the right triangle from each picture, we deduce that *the square on the hypotenuse has the same area as the squares on the other two sides* (combined). And *that's* Pythagoras's theorem.

My proof traces a path through the mathematical maze. Each node – each statement involved – leads off in all sorts of strange directions, mostly fruitless.

How did Pythagoras (don't forget, that's my name for whoever really deserves the credit) invent *his* proof?

He could have used some kind of trial-and-error method, but life is too short. On the other hand, he couldn't have just followed a systematic path. The mathematical maze is infinitely deep, so the Depth *First* algorithm would have led to a chain of ever more complex statements that were also ever less interesting. I think that Pythagoras did what all modern mathematicians do: he had some kind of 'map' of the maze in his mind. It showed what was already known, and it also had signposts pointing along what looked like interesting directions to try. And, dimly visible in the distance, was a hazy, cloud-enshrouded peak. It was either the statement he was looking for, or it was an illusion that would evaporate in the noonday sun – a cloud without an actual peak behind it.

So Pythagoras knew roughly where to start, and what direction to travel in. But, as anyone threading a maze knows, if you travel straight towards the goal you get stuck pretty quickly. The trick is to recognise the promising avenues that branch away from the obvious route. The more adept you become at doing this, the better a mathematician you will be.

How do you get that good?

By reading books like this one, of course.

JUNCTION THREE

*P*assage Two ends in a jumble of short dead-ends, turning back on itself and zigzagging through a blind corner. From here there is only one plausible exit, a narrow doorway. Just past the doorway, down at floor level, hidden from sight until you pass through, are three tiny catflaps. You bend down, and tentatively open one. You peer through.

Two eyes peer back at you. Above them are two horns.

The creature bleats.

Startled, you jump back. Then you wonder what a goat is doing behind a catflap.

Your eye is drawn to a large brass plaque on the wall. Engraved upon the plaque is a series of emotive declarations:

- YOU BLEW IT! AS A PROFESSIONAL MATHEMATICIAN, I'M VERY CONCERNED WITH THE GENERAL PUBLIC'S LACK OF MATHEMATICAL SKILLS. PLEASE HELP BY CONFESSING YOUR ERROR AND, IN THE FUTURE, BEING MORE CAREFUL.
- THERE IS ENOUGH MATHEMATICAL ILLITERACY IN THE WORLD, AND WE DON'T NEED THE WORLD'S HIGHEST IQ PROPAGATING MORE. SHAME!
- YOUR ANSWER TO THE QUESTION IS IN ERROR. BUT IF IT IS ANY CONSOLATION, MANY OF MY ACADEMIC COLLEAGUES ALSO HAVE BEEN STUMPED BY THIS PROBLEM.
- MAY I SUGGEST THAT YOU OBTAIN AND REFER TO A STANDARD TEXTBOOK ON PROBABILITY BEFORE YOU TRY TO ANSWER A QUESTION OF THIS TYPE AGAIN?
- I AM IN SHOCK THAT AFTER BEING CORRECTED BY AT LEAST THREE MATHEMATICIANS, YOU STILL DO NOT SEE YOUR MISTAKE.
- MAYBE WOMEN LOOK AT MATH PROBLEMS DIFFERENTLY THAN MEN.
- YOU'RE WRONG. BUT LOOK ON THE SERIOUS SIDE. IF ALL THOSE PH.D.S WERE WRONG, THE COUNTRY WOULD BE IN

VERY SERIOUS TROUBLE.
- HOW MANY IRATE MATHEMATICIANS ARE NEEDED TO GET YOU TO CHANGE YOUR MIND?

The emotional temperature is high, the opinions caustic and unequivocal. At least one is Politically Incorrect. A lot of clever people have become extremely incensed about something.

You enter Passage Three with more than a little trepidation, wondering what terrible fate lurks within its narrow confines.

Passage Three

MARILYN AND THE GOATS

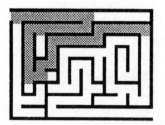

A few years ago there was quite a fuss in American newspapers about a problem involving probabilities. They called it the Monty Hall Problem, after a well-known TV game show host.

The fuss was started, entirely innocently, by Marilyn vos Savant, who is listed in the *Guinness Book of World Records* as having the highest IQ ever recorded. She writes a regular column called 'Ask Marilyn', which is syndicated in several hundred American newspapers. Marilyn described a game show contestant who has to choose one of three doors. Behind one is the major prize – a car – and behind the other two are goats. The probabilities of the car being behind any given door are equal – one in three. But after the contestant has chosen a door, the host (who knows where the car is) points to another door, different from the one chosen by the contestant, and opens it to show that there is a goat behind it.

He can always do this, you understand: since there is only one car, at least one of the two unchosen doors conceals a goat. Sometimes they *both* do (when the contestant has chosen the car).

Now comes the big question. Does the contestant's chance of winning the car improve if they now change their mind and switch to the 'third' door – the one they *didn't* choose to begin with, and which the host has *not* opened?

To be honest, I thought long and hard before deciding to include the

Monty Hall problem in *The Magical Maze*, because it caused Marilyn nothing but grief. When I wrote about the problem some years ago, I got the biggest mailbag I've ever received – mainly telling me that I was wrong. And when I mentioned the problem on a radio show a few years later, much the same happened. Fortunately I was in mid-air half-way between Singapore and Australia when the show was broadcast, and by the time I could be contacted the hoo-ha had pretty much died down.

Why all the fuss?

A lot of people have a simple and powerful intuition about the problem. This intuition is *wrong*, but it's not at all obvious *why* it is wrong.

Here's that intuition. It's hard to see how the host's information could possibly make any difference. The car is just as likely to be behind the door that you first chose as it is to be behind that one remaining door. Since the odds are fifty–fifty either way, why swap?

Marilyn explained exactly what's amiss with this seductive line of reasoning, showing – in detail and at length – that the probability of winning the car is *doubled* if you swap. It improves from one chance in three, probability 1/3, to two chances in three, probability 2/3. This result is so counter-intuitive that many people find it hard to accept the verdict of the mathematics. 'But you're choosing between two doors. Each is equally likely. Therefore it can't be advantageous to swap.' The fallacy is that the probabilities are *not* equally likely: the two doors under discussion have been selected by a process that is *based on your original choice*.

This biases the probabilities. If your first choice was right, you'll lose the car if you swap; but if it was wrong, then a swap will secure the car. The upshot is this: the probability of the car being behind the door that you did not choose and which the host did not open is 2/3, not 1/2. The probability of the car being behind the door that you originally chose is 1/3 – *just what it was when you made that choice*.

The emotive declarations on the plaque at the entrance to Passage Three of the magical maze are a small selection of the responses from Marilyn's readers – all at universities, colleges, or research institutes. She explained her answer a second time, but the deluge of irate mail just got worse. She received 'thousands of letters, nearly all insisting that I'm wrong, including one from the Deputy Director of the Center for Defense Information and another from a research statistician at the National Institutes of Health'. In all, 92% of letters from the general public disagreed with her and sided with the contestant; the same was true of 65% of the letters from universities.

However, mathematical questions are not settled by democratic votes – and a good job too, because in this particular instance Marilyn was right and the deluge of critics was wrong.

To be fair, the responses were pretty certainly biased against her. I am sure that there must have been *some* mathematicians, statisticians, and puzzle freaks who knew the correct answer. But who would bother to write to a columnist to say 'Hey, Marilyn, great! You got it right!'? You see, the whole thing is a terrible old chestnut. If the person who recommended that Marilyn consult a textbook on probability had done so himself, he would almost certainly have found that the book discussed the Monty Hall Problem – though not under that name – and that it agreed with Marilyn.

WHAT IS PROBABILITY?

We're going to work our way up to the Monty Hall Problem by way of several simpler puzzles and paradoxes involving the elusive notion of probability. Along the way I want to convince you that human intuition for probability is unusually poor – far poorer than our intuition for arithmetic or geometry. This really is a case where cultivating a mathematical mind pays dividends – perhaps literally, since a key area where probabilities turn up is gambling. But you don't have to agree with me on that point to appreciate that probabilities are not especially intuitive.

For a start, it's not terribly clear what probability *is*.

Mathematicians can tell you a technical definition, about 'measures on a sample space', which has been arrived at after centuries of concentrated thought. Like most such products of the mathematical mentality, this definition is so elegant and abstract that its connection with our intuitive feelings about probability is apparent only to an expert. So I'm going to take a more freewheeling attitude, giving you a basic feel for probability without trying to convey the content of the mathematicians' definition.

In this spirit, probabilities are numbers that tell us how likely certain events are. An event whose probability is 1/2, for instance, is equally likely to happen or not to happen. An example is 'tossing heads with a fair coin'. An event with probability 1/6 is one that will happen, in the long run, one time in six – such as 'throwing a six with an unbiased die'. An event with probability 1/13,983,816 is one that will happen, in the long run, one time in 13,983,816 – such as 'winning the jackpot in the UK National Lottery'.

There are at least two tricky features of this description. The first one is the meaning of 'in the long run', and I'll come back to that when we've got some basic points of understanding sorted out. The second, which caused mathematicians a lot of worry until they realised it's a total red herring, focuses on words like 'fair' and 'unbiased'. What do I mean by a fair coin? One for which 'heads' or 'tails' are equally likely, right? But then my definition is circular – I define the probability to be 1/2 if the coin is fair, and I define it to be fair if the probability is 1/2. The answer to such worries is not to use the above description as a *definition*. Instead of telling us what probabilities *are*, it tells us what they *do*. Terms like 'fair' are a reminder that certain modelling assumptions are being made. When I say 'fair coin' I am setting up a convention about how we handle the mathematics, not talking about any real coin. Whether the pound coin in your pocket is fair or not is a matter for physical experiment, not mathematical definition. (My guess is that it's *not*, actually, because the design on one side differs from that on the other. Since the design is incised, the coin's centre of mass is probably not quite in the middle. This should, in the long run, bias it slightly towards one face rather than the other – my guess is that 'heads' is slightly more common, since the design on the tails side looks a bit heavier and will therefore cause the coin to land tails down a little more often than tails up.) 'Fair coin' is a mathematical idealisation of reality, just as 'circle', 'line', 'sphere', and so on are. The question is whether it's a useful idealisation, not whether it is 'true' – and here the verdict of much experiment is 'yes'.

So that's all right, then.

Aside from fine details of the design, there's no obvious reason why a coin should fall one side or the other when tossed, so the assumption that a real coin can sensibly be modelled by a mathematician's ideal fair coin is plausible. Similarly, the six faces of a well-made die don't differ significantly (again, apart from a slight imbalance caused by the dots being indented), so we would expect such a die to be sensibly modelled by a 'fair die' in which each face has probability 1/6 of coming to the top. These statements aren't *proofs* – proofs apply to mathematical objects, not to real-world realisations of them – but they are excellent motivation for the modelling assumptions.

Probabilities can be expressed in several ways. Instead of saying that the probability that a coin lands 'heads' is 1/2, I can say that it is 0.5. Or I can say there's a 50% probability of it landing heads. Or that the chance of getting heads is one in two. These are all different ways of saying the same thing. Gamblers have yet another way to say this: the odds are even. 'Odds' here has nothing to do with even and odd numbers. The odds are what you will win, in addition to getting your stake back, should your guess prove to be right. If I bet £1 on a horse at odds of 7 : 1, I get back £7 in winnings *plus* my £1 stake. The bookmaker will break even in the long run if the probability of that horse winning is 1/8 (not 1/7). Odds are 'even' when they are 1 : 1 – win £1 and get back your original £1. The probability here is 1/2.

THE LAW OF AVERAGES

The basis of many misconceptions about probability is a belief in something usually referred to as 'the law of averages', which alleges that any unevenness in random events gets ironed out in the long run. For example, if a tossed coin keeps coming up heads, then it is widely believed that at some stage there will be a predominance of tails to balance things out.

Is that true? Take ten coins and 'prepare' them, so that – if the 'law of averages' is right – they are more likely to land tails. Toss each coin repeatedly until you get a run of (say) four heads, HHHH. Does such a run somehow improve the odds on getting tails on the next throw of that coin? Toss each of the ten 'pre-prepared' coins once, and see how many tails you get. Most of you won't find any noticeable difference in the numbers of tails thrown with a 'prepared' coin and from any other coin. (A few will get results wildly out of line – those are statistical fluctuations, and they should be rare.) Coins have no memory, there is no way for their past history to influence their future. If there was, Mike Atherton would carry a pocket full of pre-prepared coins for the toss that decides who opens the batting in a cricket match. Cunningly they would look exactly like any other coin, but they would be charged to the eyeballs with un-fulfilled probability, a dead cert for turning up tails.

Come on. Mathematics may be magical, but it's not *that* magical.

Convinced? Good. However, now I have to point out that there is an interesting sense in which things *do* tend to balance out in the long run!

Suppose you toss a fair coin lots of times, and plot the excess of H's over T's by drawing a graph of the difference at each toss. You can think of this as a curve that moves one step upwards for each H and one down for each T. So the sequence of tosses

TTTTHTHHHHHHHTTTHTTTH

produces the graph of Figure 22. Such a graph is called a *random walk*.

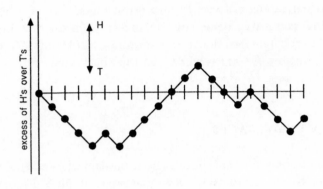

Figure 22 A short random walk.

For this picture, the numbers do in fact balance out pretty often, but that's a misleading effect created by using just a few tosses. Figure 23 shows a graph of a typical random walk – 10,000 tosses. In this particular instance, heads spend a lot of time in the lead. This sort of wildly unbalanced behaviour is entirely normal – in fact, random walk theory shows that if you make a series of 10,000 coin tosses, then one side will lead for at least 9,930 tosses (and the other for 70 or fewer) in about *one trial in ten*.

However, random walk theory also tells us that the chance that the balance *never* returns to zero – that is, that H stays in the lead *for ever* – is 0. This is the sense in which the 'law of averages' is true. If you wait long

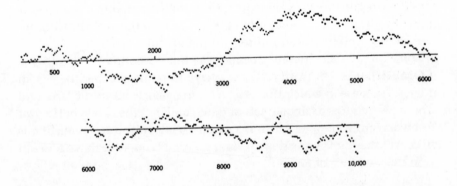

Figure 23 A long random walk: heads spends a lot of time in the lead. This is entirely typical.

enough, then almost surely the numbers of heads and tails will even out. But this fact carries no implications about improving your chances of winning, if you're betting on whether H or T turns up. The probabilities are unchanged, and you don't know *how* long the 'long run' is going to be. Usually it is very long indeed.

For instance, suppose that you toss a coin 100 times, and at that stage you have 55 H's and 45 T's – an imbalance of 10 in favour of H's. Random walk theory says that if you wait long enough, then the balance will correct itself.

Isn't that the 'law of averages'?

No. Not as that 'law' is normally interpreted. If you choose a length in advance – say a million tosses – then random walk theory tells us that on average those million tosses are completely unaffected by the imbalance. If you made huge numbers of experiments with one million extra tosses, then on average you would get 500,055 H's and 500,045 T's in the combined sequence of 1,000,100 throws. On average, imbalances *persist*. However, the proportion of times that H is thrown then changes from $55/100 = 0.55$ to $500055/1000100 = 0.500005$, which is a lot closer to 1/2.

The 'law of averages' asserts itself not by removing imbalances, but by swamping them.

Random walk theory tells us that if you wait long enough – on average, *infinitely* long – then eventually the numbers will balance out. If you stop at that very instant, then you may imagine that your intuition about a 'law of averages' is justified. But you're cheating: you stopped when you got the answer you wanted. Random walk theory also tells us that if you carry on for long enough, you will reach a situation where the number of H's is a billion more than the number of T's.

If you stopped there, your experiment would support a very different intuition.

COMBINING PROBABILITIES

The raw ingredients of probability theory are called 'outcomes' and 'events'. The image is that of repeatedly carrying out some experiment, or 'trial' – tossing a coin, say, or rolling a die. The possible results of the experiment are 'outcomes'. For the coin, the two outcomes are 'heads' and 'tails'. For the die, the outcomes are '1', '2', '3', '4', '5', and '6'.

Events are more complicated: they are *lists* of outcomes. If the list had just one outcome in it, then there's no great difference between an outcome and an event – but there is if the list has more (or less) than one outcome in it. People who bet on horses do not just bet that a horse called, say, Marilyn's IQ will win. They may bet 'each way' – that it will win, or come second, or come third. That's a list of outcomes, not a single outcome. It's a typical event, in the mathematical sense.

When you toss a coin, 'heads' is an event, and so is 'tails'. The only other lists are 'heads and tails' and 'neither heads nor tails'. The first doesn't mean that the coin comes up both heads and tails! It means that you win if the coin comes up heads *and* you win when it comes up tails. The probability of this event is 1 – complete certainty. The second is a fancy way of saying 'nothing happens at all', an event whose probability is zero, meaning that it doesn't happen.

Here I'm ignoring 'lands on edge' – it's relevant to a real coin, and I've even seen it happen once, but the usual idealisation of a coin excludes it from consideration. This is because we want to use the image of a coin to understand a mathematical concept, not the other way round: a good model of a physical coin might well include 'edge' with a small probability. But such a 'coin' would be more complicated to analyse.

When you roll a die, events are things like '6', '2', 'an even number', 'an odd number different from 3', and so on. I repeat: events are not individual outcomes, but *lists* of outcomes. Of course, that list *may* contain just one item, so you can think of any given outcome as an event in its own right. A bet on Marilyn's IQ to win is just such an event.

The mathematics of probability, for our purposes, boils down to this: the probability of an event is the number of ways that it can happen, divided by the total number of outcomes – provided we set the calculation up so that all of those outcomes are equally likely.

For example, consider a fair die. The total number of outcomes is six, all equally likely since we're assuming the die to be fair. The event 'an even number' can happen in three ways – 2, 4, or 6. So the probability that you throw an even number is 3/6, which equals 1/2. The event 'an odd number different from three' can happen in two ways – throwing 1 or 5. So its probability is 2/6, or 1/3. From this point of view probabilities boil down to counting things. There's a more sophisticated approach for problems where counting things doesn't work – like 'the probability that you will be hit by a meteorite' – but that comes under the abstract definition of probability, and it would lead us *much* too far astray.

To save ink, let's abbreviate 'the probability of event E' to

$$P(E).$$

For example, instead of saying that 'the probabilty of throwing 6 with a fair die is 1/6', I can write

$$P(6) = 1/6.$$

Using this notation, I'll now express some basic laws of probability.

There are two extreme cases. If E is the event 'any of the possible outcomes happens', then

$$P(E) = 1$$

and E is *certain*. If E is the event 'none of the possible outcomes happens', then

$$P(E) = 0$$

and E is *impossible*.

The most important thing you need to know about probabilities is how to combine the probabilities of simple events to get the probabilities of complicated ones. There are three main operations of this kind: 'not', 'or', and 'and'.

With one throw of a fair die, what is the probability of 'not 6'? (In our new notation this is P(not 6).) Well, there are six outcomes. Five of those are 'not 6', namely '1', '2', '3', '4', '5'. So P(not 6) = 5/6. Now, there's a quicker way to see that this is the answer. There is exactly one outcome leading to '6', namely '6' itself. So the number of outcomes leading to 'not 6' must be $6 - 1 = 5$. We don't need to list them to see this. The probability of 'not 6' is therefore $(6 - 1)/6$, which we can write as $1 - 1/6$. In short,

$$P(\text{not-}6) = 1 - P(6).$$

The reasoning is completely general, and it shows that

$$P(\text{not-}E) = 1 - P(E).$$

The total number of outcomes is the number listed in E, plus the number *not* listed in E. As proportions of the total number, these two quantities add up to 1. So each is 1 minus the other.

That's 'not'. What about 'or'? Think about the event 'the outcome of tossing the die is either an even number or it is an odd number not equal to three'. Here we have two events:

$$E = \text{'even number'},$$
$$F = \text{'odd number not equal to three'}.$$

Event E can happen in three ways – 2, 4, 6. Event F can happen in two ways – 1, 5. So either E or F can happen in five ways – 1, 2, 4, 5, 6. So we have

$$P(E) = 3/6 = 1/2,$$
$$P(F) = 2/6 = 1/3,$$
$$P(E \text{ or } F) = 5/6.$$

Here

$$P(E \text{ or } F) = P(E) + P(F).$$

Is this coincidence, or is there a general rule here? Well, we get E or F by putting together the two lists for E and F separately, so of course the number of outcomes in E or F is the sum of those in E and in F.

Of course? What about this:

$$E = \text{'even number'},$$
$$F = \text{'multiple of three'}.$$

Then E contains the outcomes 2, 4, 6 and

$$P(E) = 3/6 = 1/2$$

again. Moreover, F contains the two outcomes 3 and 6, so

$$P(F) = 2/6 = 1/3$$

again. But this time E or F contains only four outcomes: 2, 3, 4, 6. Why not five? Because one outcome, here 6, is included in *both* lists.

Well, right, sure. The number of items in the combined list is only the sum of the numbers of items in each component list if the two lists *don't overlap*. Provided this happens, then clearly

$$P(E \text{ or } F) = P(E) + P(F).$$

But this is true only for non-overlapping events.

The final rule of probabilities applies when we perform two experiments in succession, and want to know the probability of an event E occurring on the first trial *and* an event F occurring on the second. For instance, we could roll a die and then toss a coin. Suppose E is the event 'even number' and F is 'heads'. Then 'E and F' is the event 'an even number on the die *and* heads on the coin'.

We already know that

$$P(E) = 3/6 = 1/2,$$
$$P(F) = 1/2.$$

But what is P(E and F)?

Let's list the possible outcomes for the twofold trail (Table 8).

Table 8

die	coin	E and F?
1	H	
2	H	yes
3	H	
4	H	yes
5	H	
6	H	yes
1	T	
2	T	
3	T	
4	T	
5	T	
6	T	

There are twelve outcomes altogether: 6×2, the product of the number of outcomes for the first trial and that for the second. Only three of them, marked 'yes' in the third column, give a combined event in E and F. So here we have

$$P(E) = 3/6 = 1/2,$$
$$P(F) = 1/2,$$
$$P(E \text{ and } F) = 3/12 = 1/4.$$

Clearly,

$$P(E \text{ and } F) = P(E) \times P(F)$$

in this case – and this is the general rule for *any* two events in a row, provided the second trial is *independent* of the first, meaning that the result of the first trial doesn't alter that of the second. (The coin doesn't 'know' what the die did.)

Let's call these three rules the Not Rule, the Or Rule, and the And Rule. Here they are again, to summarise:

Not Rule $P(\text{not } E) = 1 - P(E).$

Or Rule $P(E \text{ or } F) = P(E) + P(F)$ for non-overlapping events.

And Rule $P(E \text{ and } F) = P(E) \times P(F)$ for independent events.

Now we're in business!

A MAZE OF CHANCES

Pardon me for hammering my 'maze' metaphor, but it applies to probabilities too. A probabilistic trial, such as tossing a coin, can be represented as a simple maze (Figure 24(a)). It has one entrance ('perform a trial') and two exits, one for the outcome 'heads' and one for the outcome 'tails'. The probabilities are numerical labels attached to the

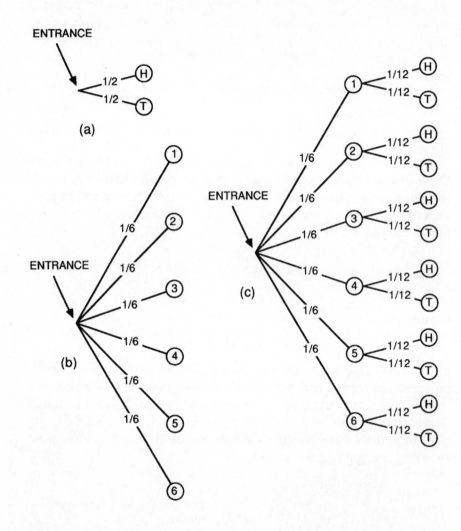

Figure 24 Probability mazes: (a) tossing a coin; (b) rolling a die; (c) rolling a die and then tossing a coin.

passages of the maze, and they tell you how likely it is that the trial will wander along that path. The same goes for rolling a die, but now there are six exits (Figure 24(b)).

When we perform two trials in succession, we combine their mazes. Suppose the experiments are 'roll a die' and then 'toss a coin', as above. Then the combined maze is as shown in Figure 24(c). The probabilities assigned to the passageways for the first experiment are exactly as before. But attached to each exit of that first maze is a copy of the second maze. By the And Rule, the probabilities attached to these new passages are not those of the corresponding maze for the second experiment, but those probabilities *multiplied* by the probability of ever getting into that section of the big maze. That is, not $1/2$ but $1/6 \times 1/2$.

So the And Rule is really a way of telling us how to label combinations of probability mazes.

If you look at this kind of probability maze, a rather different image may occur to you. Start at the entrance with a total probability of 1. The maze branches into six passages, all equally likely. The total probability correspondingly splits into six probabilities of $1/6$, one for each passage, and the probability 'flows along' the passage to its far end. Here there are two branches, both equally likely, so now the total probability of $1/6$ splits into two pieces, each half as big – value $1/12$. Those probabilities 'flow' along the passage to the exit.

You could imagine replacing the maze by a system of pipes, branching in exactly the same manner. You put one litre of water in at the entrance. It flows along the pipes, dividing according to the probabilities, so $1/6$ of a litre flows along each of the first system of pipes (Figure 25). Then the maze of pipes splits again, and $1/12$ of a litre flows along each of the pipes for the second trial.

Pour a litre of water in at the entrance, go to whichever exit corresponds to a desired outcome, and catch the water that comes out in a bucket. The amount of water in the bucket is the probability of getting that outcome.

What about an event? That's a list of outcomes. Provided you are sensible and list each of them exactly once, the list has no overlaps. The Or Rule says that we get the right answer for the probability of an event by putting buckets under all the outcomes in the list, collecting the water that comes out of each, and pouring it all into one container.

What about the Not Rule? That says that the water not collected in a bucket is the amount you start with (1 litre) minus the amount that you collect.

It all makes perfect sense.

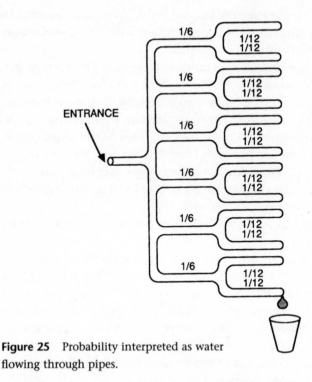

Figure 25   Probability interpreted as water
flowing through pipes.

Probabilities aren't just numbers, and they aren't just frequencies-on-average. They are also rather like a *substance* that flows, dividing according to the likelihood of various outcomes, subdividing when several trials are performed in succession, and adding together when several outcomes are combined to give an event. This is a metaphor, but an accurate and powerful one. It is, in a sense, the metaphor that mathematicians formalise when they offer a *definition* of probability. In this sense, probability behaves like volume, mass, or area. The technical term is 'measure'. The technical definition of probability is 'a measure such that various nice things happen'.

Probability is a quantity that flows through the conceptual maze of possible events, and it behaves just like water flowing through pipes.

BIRTHDAY COINCIDENCES

I said that we don't have a good intuition for probabilities. Let me prove this to you by asking for an *immediate* 'guesstimate' for the answer to the following question: 'How many people should there be in a room in order for the probability that at least two of them have the same birthday to be more than 50%?' In other words: there's a bunch of people at a party, and they start comparing birthdays. (We assume throughout that all birthdays are equally likely. If it's a meeting of the 'Born on the Fourth of July' Club, you'll get very different answers.) If there is only one person, then the probability that the same birthday occurs twice is zero. If there are 366 people, then (ignoring 29 February) the same birthday *must* occur twice – probability 1. Somewhere in between 1 and 366 is the 'break even point' where, on average, half the parties include two people or more with the same birthday. How big is such a party? 100? 150? 182? After all, there are 365 possible birthdays – 366 if we include 29 February in leap years. To keep the analysis simple, we won't do that, but it would have little effect on the answers if we did.

Would you believe 23?

It seems surprisingly small. But 23 is the right answer. Before I explain why, here's another deceptively similar question: 'How many people, in addition to yourself, should there be in a room in order for the probability that at least one of them has the same birthday *as you* to be more than 50%?'

This time the answer is 253 – not counting you. Sounds *way* too big, right? And why do two such similar questions have such different answers?

Our job is to see why the numbers turn out as they do. Along the way we shall come to understand why our 'intuition' about these numbers is badly off the mark. In fact, our intuition about probabilities often leads us astray – in many ways probability seems to be one of the areas in which human intuition is the least reliable. One consequence of this is that we tend to be unnecessarily impressed by 'coincidences' that are actually fairly likely. Another is that we adopt nonsensical strategies when betting on such things as horse races or lotteries. Our geometric intuition, based on our visual sense, is much better matched to reality. Perhaps the reason for the difference is that our evolutionary history has seldom placed us in situations where an intuitive grasp of probabilities adds much survival value. Usually a crude division into 'impossible', 'very unlikely', 'might happen', 'probably will happen', or 'certain' is enough.

Let's think the birthday questions through, instead of leaping to unwarranted conclusions. One of the most useful tricks in probability calculations is to consider the probability that the event under consideration does *not* happen. If we know this number, then the Not Rule tells us that all we have to do to get the probability that the event *does* happen is to subtract the number from 1. And it's amazing how often it's easier to calculate the probability that an event does *not* happen.

For instance, when does the event 'at least two people have the same birthday' *not* happen? When all of their birthdays are different, of course. Suppose we increase the number of people in the room, one at a time, starting with just one of them. We can calculate the probability that the new person has a different birthday from all the previous ones. We'll see how difficult it gets for this to happen. As soon as the probability drops below 50%, we know that the probability of the original event – at least two people have the *same* birthday – has risen *above* 50%.

OK, here goes. With one person only, there is no problem in making all birthdays different! So the probability here is 1.

Now we introduce the second person. There are 365 possible birthdays, but person 1 has used up one of them. That leaves 364 possibilities for person 2, if the two birthdays are to be different. So with two people in the room, the probability that their birthdays are different is 364/365.

Enter person 3. Now there are only 363 choices that keep all birthdays different, so the probability that person 3 has a birthday that differs from the other two is 363/365. By the And Rule, the combined probability that person 2's birthday differs from that of person 1, and person 3's birthday differs from those of persons 1 and 2, is

$$(364/365) \times (363/365)$$

Now we're starting to see a pattern. When person 4 comes into the room, the And Rule tells us that this probability must be multiplied by 362/365, to give

$$(364/365) \times (363/365) \times (362/365),$$

and so on. In general, after person n has entered the room, the probability that all n birthdays are different is

$$(364/365) \times (363/365) \times \ldots \times (365-n+1)/365).$$

All we have to do now is tabulate successive values of this expression, and see when it drops below 0.5 (a probability of 50%). Table 9 shows what happens.

Table 9

No. of people	probability	No. of people	probability	No. of people	probability
1	1.000	9	0.905	17	0.684
2	0.997	10	0.883	18	0.653
3	0.991	11	0.858	19	0.620
4	0.983	12	0.832	20	0.588
5	0.972	13	0.805	21	0.556
6	0.959	14	0.776	22	0.524
7	0.943	15	0.747	**23**	**0.492**
8	0.925	16	0.716	24	0.461

We see that with 22 people the probability of all birthdays being different is 0.524, slightly bigger than 0.5; but that with 23 people it is 0.492, slightly less than 0.5. Which means that when 23 people enter a room, the probability that at least two of them have the same birthday is $1 - 0.492 = 0.508$, slightly bigger than 0.5.

Only 23. Amazing. I know how to calculate it, and I still have difficulty believing it. But it's true. Try it at parties with over 23 people. Take bets. In the long run, you'll win. At big parties you'll win easily.

Why does our intuition go so badly wrong? I suspect that it is because we focus only on one aspect of the problem. Let's compare with my second question. *You* are in a room and people start to enter. How many of them must there be in order for the probability that one of them has the same birthday *as you* to be more than 50%? Is the answer 364/2, or 182 people? After all, there are 364 birthdays that differ from yours, and once half of those have been used up ...

No, as I said, it is actually 253.

Now *that* sounds much too *big*! But again, a little thought shows that I'm right. Again, we focus on the probability that the birthdays remain *different* from yours, and then subtract from 1.

Suppose, for the sake of argument, that your birthday is 24 September. The probability that the first person to enter the room has a different birthday from 24 September is 364/365. The probability that the second person to enter the room has a different birthday from 24 September is *also* 364/365. And the same goes for the third, fourth ... nth person. We're not interested in coincidence between the birthdays of the people that come in – say Fred and Gina both have birthdays on 19 May. Doesn't matter: all that counts is whether their birthday is 24 September. By repeated use of the And Rule, the probability that after n people have entered they all have different birthdays from 24 September is therefore $(364/365)^n$. The first value of n for which this number is less than $1/2$ is $n = 253$; in fact $(364/365)^{253} = 0.499$, but $(364/365)^{252}$ is slightly bigger than 0.500.

The same calculation works whatever your birthday is, of course. There's nothing special about 24 September.

Even though we tend to underestimate this number, it's much bigger than the 23 that shows up in the previous problem. I suspect that our intuition gets the two problems mixed up – as well as underestimating the second. When thinking about the first problem, we subconsciously simplify it by thinking of just one person's birthday – perhaps our own, or perhaps one of the two (supposed) people who have the same birthday – and wonder how many people are needed before there is a better than evens probability that our birthday is duplicated. But that's a very different question. The room can be full of people whose birthdays are not the same as ours; but if any two of them have the same birthday as each other, then the game is up. By focusing on just one date, we forget how difficult it is to keep missing all the others as the number of people piles up.

Why do we underestimate when it comes to our own birthday being duplicated? I think we make the opposite mistake. We think of those 364 slots that are not our birthday as filling up, one by one. But they don't have to. As the new people come in, many slots are filled more than once as their birthdays duplicate those that have already occurred. So it is *harder* to duplicate that one birthday out of 365 than we would guess.

When it comes to probabilities, there really is no substitute for working them out systematically.

BEYOND REASONABLE DOUBT

Probability is becoming an important issue in human affairs. Not just in gambling games, but in law.

Mathematics is invading the courtroom.

Juries used to be instructed to convict the accused of a crime provided they were sure 'beyond reasonable doubt' of their guilt. This instruction is somewhat qualitative: it all depends on what each juror considers to be reasonable. A future civilisation might attempt to quantify guilt by adopting a common science fiction scenario, in which the jury is replaced by the court computer. The computer weighs the evidence, calculates a probability of guilt, and terminates the trial when that probability becomes sufficiently close to 1. But today's civilisation does not have court computers, so juries are being forced to grapple with probability theory.

One reason is the increasing use of DNA evidence, a circumstance dramatised by the trial of O.J. Simpson☞. The science of DNA profiling is relatively new, so the interpretation of DNA evidence relies upon assessing probabilities. Similar problems could have arisen when conventional fingerprinting was first introduced, but lawyers were presumably less sophisticated in those days; at any rate, fingerprint evidence is no longer contested on probabilistic grounds☞.

In 1995 the mathematician and journalist Robert Matthews pointed out that an even longer-standing source of evidence in court cases *ought* to be analysed using probability theory – namely confessions☞. To Tomás de Torquemada, the first Spanish Grand Inquisitor, a confession was complete proof of guilt – even if that confession was extracted under duress, as it generally was. Toquemada specifically authorised the use of torture to obtain evidence, and is estimated to be responsible for some two thousand people being burnt at the stake on the basis of forced confessions. Modern legal practice is generally sceptical about confessions *known* to have been obtained under duress. Despite this, in the UK a whole series of high-profile terrorism convictions, hinging on confessional evidence, have in the last few years been overturned because of severe doubts that the confessions were genuine.

One of Matthews' most surprising conclusions is that there are circumstances under which the existence of a confession adds weight to the view that the accused is *innocent* rather than guilty. He calls this discovery the 'Interrogator's Fallacy'. The idea offers a general reason for mistrusting confessions in terrorist cases, unless they are supported by appropriate corroborative evidence. Whatever your views on this application, the analysis sheds some interesting light on probabilities.

The main mathematical idea required is that of conditional probability. Later, it will also illuminate the Monty Hall Problem – I haven't forgotten about that. But since the idea of conditional probability hasn't been introduced yet, I'll start with a simpler problem.

Mr and Mrs Smith tell you that they have two children, and one of them (at least) is a girl. What is the probability that the other child is a girl too? We'll assume that boys and girls are equally likely, each with a probability of 1/2. That's not quite true in the real world, but the extra verisimilitude derived by using the correct probabilities – about 52% for a boy, 48% for a girl – complicates the problem without shedding much extra light on it. We'll also assume that when the Smiths say that at least one child is a girl, they're not committing themselves in any way about whether both are. No quibbles like 'Ah, but the Smiths would never say *that* if both children were girls.' The Smiths are here for mathematical purposes, and they are being inscrutable about the status of their children.

The reflex response is that the other child is either a boy or a girl, and each is equally likely, so there is a probability of 1/2 that both children are girls.

Not so.

Let's think it through, instead of trusting our highly questionable intuition. There are four possible gender distributions: BB, BG, GB, GG – where B and G denote 'boy' and 'girl', and the order is that in which, say, the children were born. Each combination is equally likely, and so has probability 1/4 by the And Rule. In exactly three *equally likely* cases, BG, GB, GG, the family includes a girl. So the possible outcomes of the Smith's two-child 'trial' are three in number, all equally likely.

In just one of those trials, GG, is the other child also a girl. So the event 'two girls' happens in one way out of the three equally likely alternatives. In other words, the probability of the Smiths having two girls, *given* that there is at least one girl, is 1/3.

This is how *conditional* probabilities – probabilities of some event given that some other event has happened – are calculated. List all events and their probabilities, remove those in which the conditional event does not happen, look at what's left, and do your counting accordingly. And it really does correspond to the Smiths and their predicament, for the following reason. Suppose we looked at a large number of cases of people with exactly two children (no twins – idealised models, OK?). Maybe there are 4000 such families. Roughly a thousand of them will have BB, another thousand BG, another thousand GB, and the final thousand GG. Not exactly, you understand, but close – and on average, over lots of samples of such families, those are the correct proportions. Note that BG and GB really are different cases: when the families concerned included only one child, that child was B for one family but G for the other.

We are told that the Smiths have at least one girl. This puts them into the BG, GB, or GG bracket – 3000 possible families altogether. Of those three thousand, exactly one thousand have two girls. Probability: 1000/3000 = 1/3, as I claimed.

In a large number of two-child families, roughly half will consist of one boy and one girl *in some order*, one-quarter will be all-boy, and one-quarter all-girl. Of the families that include a girl, about one-third will be all-girl and two-thirds boy/girl. If you don't believe me, keep tossing two coins, where H represents a boy and T a girl, and see what happens.

So far, most people are reasonably happy with the answers I've given. Where a lot of them blow their top is if we change the set-up, subtly. Suppose that the Smiths tell you that their *eldest* child is a girl. What is the probability that the youngest is a girl too?

This time both BB and GB are eliminated by the information we are given. The possible gender distributions are BG and GG, each equally likely. The youngest is a girl only for GG, so the probability becomes 1/2. In particular, the answer is *different*. The story of the Smiths' children shows that the use of conditional probabilities involves specifying a context – and the choice of context☛ can have a strong effect on the computed probability.

To see how subtle such issues are, and why the answer depends on the assumed context, suppose that, as always, you already know that Mr and Mrs Smith have exactly two children – but nothing else about them. One day you see the children playing in their garden. One child, visible, is a girl. The other, playing with the dog, is partially hidden, and its sex is uncertain. What is the probability that the Smiths have two girls? You could argue that the question is just like the first scenario above, giving a probability of 1/3. Or you could argue that the information presented to you is 'the child not playing with the dog is a girl', like the second scenario in that it distinguishes one child from the other, so the answer is 1/2. Mr and Mrs Smith, who know that the child playing with the dog is young William, would say that the probability of two girls is 0. So who is right?

Which answer is correct depends, as I've said, on your choice of context. Have you sampled randomly from situations in which there are (in principle) many different families in which either child, randomly, plays with the dog? Or from families in which only one child – the youngest, say – ever plays with the dog? Or are you looking only at a specific family, in which case probabilities are the wrong model altogether?

The answer you get depends on which assumptions fit the scenario. *All three assumptions* could be valid models – but for different settings. The interpretation of statistical data required an understanding of the mathematics of probability *and* the context in which it is being applied. And that brings us back to the law.

THE INTERROGATOR'S FALLACY

Throughout the ages, lawyers have shamelessly abused jurors' mathematical unsophistication, either to obtain convictions of innocent people or to acquit the guilty. One example in DNA profiling – now well understood by the courts, but still regularly tried on by unscrupulous lawyers – is the 'Prosecutor's Fallacy'.

The idea of DNA profiling was invented in 1985 by Alec Jeffreys of the University of Leicester, and centres around so-called VNTR (variable number of tandem repeat) regions in the human genome. In each such region a particular DNA sequence is repeated many times. VNTR sequences vary greatly between individuals, and are widely believed to identify them uniquely. In 'multi-locus probes', standard techniques from molecular biology are used to look for matches between several different VNTR regions in two samples of DNA: one related to the crime, the other taken from the suspect. Sufficiently many matches should provide overwhelming statistical evidence that both samples came from the same person.

It is rather as if everybody carried around with them a large notebook, parts of which contained information such as their date of birth, name, address, and so on. By collecting enough such items of information, we would be able to argue that the Frederick Jones of 17, The Green who is in dock is the same person as the Frederick Jones of 17, The Green who dropped his notebook at the scene of the crime. It is not necessary (and in the case of DNA, not yet technologically feasible) to compare the entire notebook. My analogy is imperfect: we are not absolutely sure that two different people can't have the same notebook – especially as regards the VNTR regions, the items supposed to characterise them. Moreover, we can't 'read' the notebook with perfect accuracy. If it says Frederick James of 17, The Green, does that match Jones well enough to conclude that he is the criminal?

The Prosecutor's Fallacy is based on a confusion of two different probabilities. The 'match probability' answers the question, 'What is the probability that an individual's DNA will match the crime sample, given that they are innocent?' (How many innocent Frederick Joneses at 17, The Green are there, as a fraction of the entire population?) However, the question that should concern the court is, 'What is the probability that the suspect is innocent, given a DNA match?' (How many Frederick Joneses at 17, The Green are there *that had the opportunity to commit the crime*, as far as we can tell? How many of those did not commit the crime?)

If there are, say, twenty Frederick Joneses at 17, The Green out of a total population of twenty million, then the match probability is one in a million, or 0.0001% – and an unscrupulous prosecutor may be able to convince the jury that this is the probability that Jones is *not* the criminal. But the actual probability of this is 19/20, or 95%.

Conditional probabilities usually change when the order of the statements is swapped, so the two questions can have wildly different answers – as we've just seen. The source of the difference is contextual. In the first case, the individual is conceptually being placed in a population chosen for scientific convenience – say, people of the same sex, size, and ethnic grouping. In the second case, they are being placed in a less well-defined but more relevant population – those people who resemble them in the appropriate respect, DNA profile, and might reasonably have committed the crime.

The use of conditional probabilities in such circumstances is governed by the mathematical definition of conditional probability, which can be thought of as a very special case of a general theorem credited to the English probabilist Thomas Bayes. As a result, this kind of analysis is known as *Bayesian reasoning*. Let A and C be events, with probabilities $P(A)$ and $P(C)$ respectively. Write

$$P(A \text{ given } C)$$

for the probability that A happens, given that C has definitely occurred. As before, let 'A and C' denote the event 'both A and C have happened'. Then the simplest instance of Bayes's theorem tells us that

$$P(A \text{ given } C) = P(A \text{ and } C)/P(C).$$

For example, in the case of the Smith children, first scenario, we have

$$C = \text{'at least one child is a girl'},$$
$$A = \text{'the other child is a girl'},$$
$$P(C) = 3/4,$$
$$P(A \text{ and } C) = 1/4$$

(because 'A and C' is also the event 'both children are girls', or GG). Then Bayes's theorem (or in this case definition) says that the probability that the other child is a girl, given that one of them is a girl, is $(1/4)/(3/4) = 1/3$, the value we arrived at earlier. Similarly, with the second scenario, Bayes's theorem gives the answer 1/2, also as before.

For the application to confessional evidence, Matthew lets

> A = 'the accused is guilty',
>
> C = 'the accused has confessed'.

As is normal in Bayesian reasoning, he takes P(A) to be the 'prior probability' that the accused is guilty – that is, the probability of guilt as assessed from evidence obtained *before* the confession. Let not-A denote the negation of event A (namely 'the accused is innocent'). Then Matthews uses Bayes's theorem to derive the formula

$$P(A \text{ given } C) = p/[p + r(1 - p)],$$

where to keep the algebra simple we write

$$p = P(A)$$

and

$$r = P(C \text{ given not-}A)/P(C \text{ given } A),$$

which we call the 'confession ratio'. Here P(C given not-A) is the probability of an innocent person confessing, and P(C given A) is that of a guilty person confessing. Therefore the confession ratio is less than 1 if an innocent person is less likely to confess than a guilty one, but it is greater than 1 if an innocent person is more likely to confess than a guilty one.

If the confession is to increase the probability of guilt, then we want P(A given C) to be larger than P(A), which equals p. Therefore we need $p/[p + r(1 - p)] > p$, which some simple algebra boils down to $r < 1$. That is: the existence of a confession *increases* the probability of guilt if and only if an innocent person is *less* likely to confess than a guilty one.

This actually sounds reasonable, if you think about it. But the implication is less intuitive: sometimes the existence of a confession may reduce the probability of guilt. In fact this will occur precisely when an innocent person is *more* likely to confess than a guilty one. Could that ever happen?

In terrorist cases, the answer is 'conceivably, yes'. Psychological profiles indicate that individuals who are more suggestible, or more compliant, are more likely to confess under interrogation. These descriptions seldom apply to a hardened terrorist, who will be trained to resist interrogation techniques. An innocent, bewildered person, with no training, subjected to extreme verbal threats, may well confess merely because they are at their wits' end and will say anything to get the interrogation to stop. It is plausible that this is what happened when securing the convictions that have now been reversed in UK courts.

Bayesian analysis demonstrates some other counter-intuitive features of evidence, as well. For example, suppose that initial evidence of guilt (X) is followed by supplementary evidence of guilt (Y). A jury will almost always assume that the probability of guilt has now gone up. But probabilities of guilt do not just accumulate in this manner. In fact, the new evidence increases the probability of guilt *only* if the probability of the new evidence, given the old evidence and the accused being guilty, exceeds the probability of the new evidence, given the old evidence and the accused being innocent.

When the prosecution case depends on a confession, two quite different things may happen. In the first, X is the confession and Y is evidence found as a result of the confession – for example, discovery of the body where the accused said it would be. In this case, an innocent person is unlikely to provide such information, and Bayesian considerations show that the probability of guilt is increased. So corroborative evidence that *depends* on the confession being genuine increases the likelihood of guilt.

On the other hand, X might be the discovery of the body and Y a subsequent confession. In this case the evidence provided by the body does not depend on the confession, and so cannot corroborate it. Nevertheless, there is no 'Body-finder's Fallacy' analogous to the Interrogator's Fallacy, because it is hard to argue that an innocent person is more likely to confess than a guilty one just because they know that a body has been discovered.

Of course, it would be silly to suggest that every potential juror should take – and pass – a course in Bayesian inference, but it seems entirely feasible that a judge could direct them on simple principles such as those pointed out by Matthews. Moreover, the *same* principles apply to DNA profiling, but in circumstances that are much more intuitive to jurors and are not obscured by fancy technology. A quick review of the Interrogator's Fallacy could be an excellent way to discourage lawyers from making fallacious claims about DNA evidence.

MARILYN AND THE GOATS

And now for our climax: Marilyn and the Monty Hall Problem. It's been so long since I told you about it that I'd better remind us both.

A game show contestant has to choose one of three doors. Behind one is a car, and behind the others are goats. The probability of the car being behind any given door is one in three. After the contestant has chosen a door, the host – who knows where the car is – points to another door, different from the one chosen, and opens it to show that there is a goat behind it. Does the contestant improve their probability of winning the car if they now change their mind and switch to the 'third' door – the one they didn't choose to begin with and which the host has not opened?

According to Marilyn, they definitely should: this strategy doubles the probability of winning. But is she right? To help you decide, the arguments for both sides will now be laid out in an imaginary conversation between the game show host and the contestant. Call the doors 1, 2, and 3.

HOST: You guess a door, and then we'll see what else I can safely reveal.

CONTESTANT: If you insist. I guess ... that ... (*ooohhh* ...) the car is behind door 1.

HOST: Excellent. I can now reveal that there is a goat behind door 2.

CONTESTANT: I don't see how that helps.

HOST: I thought the information might persuade you to change your mind and choose door 3 instead of 1.

CONTESTANT: Oh, come on! That's idiotic! After all, there's got to be a goat behind at least one of the doors that I didn't choose! How can it help me to choose the right door if all you do is tell me *which one* the goat is behind? What I need to know is which door the *car* is behind!

HOST: I can't show you that. All I'm allowed to show you is a door, different from the one you have chosen, that has a goat behind it. [*Pauses.*] So you're sticking to your choice, then?

CONTESTANT: Why not? All you've done is cut down the possibilities. I now know that either the car is behind the middle door or it's behind the right-hand door. The probability that it's behind the door I chose, the left-hand one, is fifty-fifty. So there's no advantage in changing.

HOST: Have it your own way. Only ...

CONTESTANT: [*Setting off towards the middle door.*] Only *what*?

HOST: Only you'd have twice the probability of being right if you changed your mind.

CONTESTANT: That's crazy!

HOST: Look, before we start making any choices or changes or anything like that, the probability that the car is behind any given door is one in three, yes?

CONTESTANT: Yes.

HOST: Good. So the probability that door 1 is correct is 1/3, whereas the probability that it's wrong is 2/3. OK?

CONTESTANT: Of course!

HOST: Let's follow it through, then. If the car is behind your initial choice, door 1, and you swap to door 3, then you get a goat. If you're *wrong*, however, and there's a goat behind door 1, then because I've kindly eliminated the other goat by showing you what's behind door 2, the car *must* be behind door 3. That is, if you're *right* and swap, you get a goat; but if you're *wrong* and swap, you get the car.

CONTESTANT: Yes, yes! So what?

HOST: Thing is, we've just agreed that your probability of being right is 1/3, but your probability of being wrong is 2/3. So your probability of winning the car is 2/3 if you swap, but only 1/3 if you don't.

CONTESTANT: [*Dumbfounded.*] But the probabilities *must* be equal! You've eliminated one door, so I've got two to choose from! Each is equally likely to be right!

HOST: [*Mutters under his breath.*] That wasn't the order we did it in.

So who's right? Let's be absolutely specific about the procedure:

- The car is definitely behind one of the doors. The probability of it being behind any particular door is 1/3.
- The host knows where the car is, but doesn't tell that to the contestant (of course).
- *First* the contestant chooses a door.
- *Then* the host points to one of the other two doors, one that he *knows* conceals a goat, and tells him or her (truthfully) that there's a goat behind it.
- *After that*, the contestant is offered the opportunity to change.

One useful trick for navigating the magical maze is to compile experimental evidence. Such evidence proves nothing, but suggests many things – and it can often set our intuition back on the right path when it has gone astray. Computers are great gadgets for carrying out mathematical experiments. Computer buffs will find it easy to write a computer program to simulate the problem, and to run it a large number of times, counting how often a contestant succeeds and how often they fail. If you enjoy programming, you might like to try this before reading any further.

I'm going to describe something very similar, but instead of using a computer I'll use a standard table of random numbers, namely the *Cambridge Elementary Statistical Tables*. That should avoid any probability of bias, and in principle you'll be able to check my calculations. Here's the method. Only the digits 1, 2, and 3 will be used, one for each door. I'll call these 'acceptable'. I run through the table of random numbers, looking at acceptable digits in the order in which they are printed.

1. The first acceptable digit A determines which door the car is Actually behind.
2. The next acceptable digit C determines the Contestant's guess.
3. The next acceptable digit D that is Different from the first two determines what the host tells the contestant. (The host must choose a door with a goat, so he can't use A; and he has to choose a different door from the one the contestant has chosen, so he can't choose C. Note that A and C may be equal: in this case the host may choose either of the doors with goats behind them.)
4. Assuming the contestant doesn't swap, count whether or not their guess is right.
5. For comparison, assume they take the host's advice, and swap to the unique door that differs from C and D. Count how often their amended guess is right.
6. Move on to the next acceptable digit and repeat the whole procedure.

The verdict of the statistical tables is shown in Table 10. If the contestant's strategy is *never* to swap, they are right 6 times and wrong 14

Table 10

The car's door A	Contestant's guess C	Host's selection D	Never swap	Always swap
2	1	3	wrong	right
1	3	2	wrong	right
1	1	2	right	wrong
2	2	3	right	wrong
1	3	2	wrong	right
3	2	1	wrong	right
3	1	2	wrong	right
2	1	3	wrong	right
2	2	3	right	wrong
2	1	3	wrong	right
3	3	1	right	wrong
2	3	1	wrong	right
2	1	3	wrong	right
3	1	2	wrong	right
3	3	1	right	wrong
2	3	1	wrong	right
2	2	3	right	wrong
1	3	2	wrong	right
2	1	3	wrong	right
1	2	3	wrong	right

times. Using the host's recommended strategy, *always* to swap, they are right 14 times and wrong 6 times.

Look, sorry, the host must clearly be right. *It's best to swap!*

You may feel that 20 tries is too small to be convincing. I warn you, you're grasping at straws – but let's try a more substantial trial. I'll report my results: you may doubt their veracity if you wish. Write your own simulation if you must, but do be careful to follow the prescribed sequence of events. If you build in short cuts of your own, you may end up begging the question.

I ran 100,000 trials on a computer. The contestant's strategy gave the correct answer in 33,498 trials, but the wrong answer in 66,502. With the host's recommended strategy, the numbers were exactly the other way round. The corresponding probabilities of 0.33498 and 0.66502 are convincingly close to the host's claimed values of 1/3 and 2/3.

Marilyn was *right*.

But …

Oh, no! If *that* doesn't convince you, what else can I try? OK, here's one of the several different explanations that Marilyn gave after her first mailbagging. Suppose (for the sake of argument) that you choose door 1. (You can check that the same kind of thing happens if you choose door 2 or 3. Anyway, what's in a number?) There are three possibilities. Each is equally likely because the probability of the car being in any particular door is one in three. An <u>underline</u> shows which door (or doors) the contestant can point to.

Contestant's strategy: never swap

door 1	door 2	door 3	
car	goat	goat	win
goat	car	goat	lose
goat	goat	car	lose

The host's recommended strategy: always swap

door 1	door 2	door 3	
car	goat	goat	lose
goat	car	goat	win
goat	goat	car	win

There's no escaping it: the contestant's strategy wins only one time out of three, whereas the host's wins two times out of three.

I reckon that our intuition gets badly confused because we are really dealing in *conditional* probabilities: the probability of something happening *given that something else already has*. We've seen that conditional probabilities are not especially intuitive. Conditional probabilities arise here because the host's choice of door *depends on what the contestant has chosen*. If the contestant has chosen correctly, then the host has a free choice of the other two doors; but if the contestant is wrong – which is twice as likely – then the host has *only one choice*, one that gives the entire game away.

This dependence of the choices is what makes the order of events matter, and is why conditional probabilities come into play. The argument that with only two doors left, the car is equally likely to be in either of them, is correct *if the host chooses first*; but wrong if the host's choice has to depend on what the contestant has already chosen. (He can't, for example, open the *contestant's* chosen door to reveal a goat – even if there's one behind it.) It's one occasion when the order in which the choices are made matters. This often happens with conditional probability, as we've already seen for the Prosecutor's Fallacy in DNA testing and the Interrogator's Fallacy with regard to confessions.

Whether you agree with me or not, *please, don't write in!* In Marilyn's case things got totally out of hand. There were stacks of mail, phone calls, fax messages, wild accusations. After a few days, though, there were also some embarrassed retractions. I doubt that the perpetrator of sexist sarcasm was among them: bigots seldom recognise when they're wrong. But if the mathematician who advised her to 'refer to a standard textbook on probability' had done so himself, and turned to the chapter on conditional probability, he might very well have come across a discussion of precisely this problem. It's an old chestnut – a puzzle that every probabilist should have run into, in one form or another, during their career. It's in Joseph Bertrand's *Calcul des probabilités* of 1889, for instance. Indeed, it is usually known as Bertrand's Box Paradox, and as Eugene Northrop observes in his *Riddles in Mathematics*☞, it 'has been used as an illustrative example in almost every subsequent textbook'. A version involving three prisoners and a prison governor is described in detail by Martin Gardner in *More Mathematical Puzzles and Diversions*☞. So not only did many of Marilyn's critics let their sense of outrage run away with their good taste, they didn't do their homework as professionals. 'If all those Ph.D.s were wrong, the country would be in very serious trouble.'

Hmmmm ...

JUNCTION FOUR

*T*he fourth passage in the Magical Maze is lined with mirrors. There are mirrors that reflect your face and body, turning left into right. There are mirrors that do not: your reflection looks normal, apart from some peculiarities with the face which you can't quite pin down – but when you try to scratch your nose, the movement in the mirror doesn't match, and your hand doesn't work properly.

It is humbling to discover that when the mirror doesn't work, you can't put your finger on your own nose.

Other mirrors are like those in fairgrounds. One gives you a fat tummy, another gives you a pointed head. One makes you seem very tall, another squashes you into the shape of a hamburger.

And some are quite unlike any found in fairgrounds – or anywhere else. There is a mirror that turns your smile upside-down. You smile, and your image frowns. You frown, and it smiles. You look serious – and so does your image. There is a mirror that appears absolutely normal but makes you look like a teapot. And as for this one – whoops! You can walk through to the far side!

This is such a surprise that you back out again, and quickly count your fingers and ears to make sure they're still all there.

You take off your left shoe, tie it to the end of a stick, and push it through the mirror. When you pull it out, it has become a right shoe. You hastily push it back in, and this time it comes out as a left shoe again.

You dimly recall hearing that many of the important molecules in your body are left-handed, and it dawns on you that since you yourself have passed through the mirror, they may well have become right-handed like the shoe – in which case, you will starve to death on ordinary food. A glance at your wristwatch confirms your fears. The numbers are all reflected, they run anticlockwise, and the watch is on the wrong wrist.

There's nothing for it: you have to brave the mirror a second time.

This time, when you come out, everything about the watch seems normal. You breathe a sigh of relief.

Then you start to wonder why your watch looked wrong. Was it your body and its accoutrements that flipped – or was it your brain?

And why wasn't it both?

You stare at the wall of the passage, and your own face stares back. Behind it, reflected from the mirror behind you, is the back of your own head. Beyond it, just visible over your image's shoulder, is your face. A million you's stretch off into the distance, becoming smaller and smaller the farther away they are. You can see no end to them. It is an awe-inspiring experience.

Then one of the teapots starts to climb out of its mirror, and you run off down the passage as fast as your heels – left and right – can carry you.

Passage Four

THE SLIME MOULD SAGA

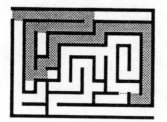

One of the most fascinating animals on the face of the Earth is – what do you think? The tiger? The sea-lion? The humpback whale? The chameleon?

Here's my candidate: slime mould.

I admit that it's not the kind of thing you expect to turn up in a public opinion poll on the fascination of animals. In fact it doesn't even sound like an animal. But it is an animal, a kind of amoeba, and it's most definitely fascinating. Because somehow this tiny, unintelligent creature manages to create the most spectacular patterns. Slime mould builds beautiful spirals (Figure 26). The same kinds of spiral arise in chemistry and in the electric impulses that keep your heart beating – and they arise for the same underlying reason. The spirals are an example of pattern formation in the natural world, and the patterns boil down – as do all patterns, really – to mathematics.

First, I'm going to alert you to the amazing abilities of the humble slime mould. Then I'm going to take a look at the mathematical theory of patterns. Finally, we'll come back to the spirals and use the mathematics to explain where the spirals come from.

SLIMY SPIRALS
Although we usually reserve the word 'animal' for something furry with

Figure 26   The elegant spirals of slime mould amoebas.

legs, in biology an animal is pretty much anything that's not a plant. An amoeba isn't a plant – it doesn't use chlorophyll to get its energy – so it's an animal. The name 'slime mould' may make us think of fungi, but slime mould is a colony of amoebas. The particular species of slime mould that I'm going to tell you about has the Latin name *Dictyostelium discoideum*, if you want to know.

The life cycle of slime mould begins – inasmuch as any cycle begins – with a spore about ten micrometres across, a dried-out amoeba that can be blown on the winds until it finds somewhere nice and moist. It then turns back into a *bona fide* amoeba, a tiny single-celled creature, and starts to do what amoebas do best, which is finding food and reproducing by splitting into two more or less identical amoebas whenever it gets big enough. You don't need to be a mathematician to see that pretty soon there'll be a lot of amoebas around. When conditions become sufficiently crowded, and it's time for the amoebas to spread themselves around more, they do something very strange indeed. Instead of just wandering off individually and dispersing, they split into small patches. All the amoebas in one patch crowd in towards a 'meeting place' somewhere near the middle.

As the crowd of amoebas makes it way in towards its common

destination, elegant spirals appear. We see the patterns because the light bounces off a crowded region of amoebas differently from a thinly populated region, so what the spirals represent is a kind of 'wave pattern' for the population density. The more amoebas there are in a given region, the darker that region gets.

If you watch for long enough, you'll discover that the spiral patterns slowly *rotate*.

This is amazing enough, but it's just the beginning. As time passes, the crowd of amoebas flowing towards the meeting point becomes denser. The spirals get more and more tightly wound – and then start to break up into 'streaming patterns' that look like roots or branches. The streams thicken, the gaps between different patches of amoebas become more clearly defined. As more and more amoebas try to get to the same place, they pile up in a heap, which eventually forms what biologists call a 'pseudoplasmodium' – more familiarly known as a 'slug'.

It's called that because it looks like one.

But the slug isn't an organism – it's still a colony. Despite that, it can move about as if it were a single organism, and it does. It heads off across the ground looking for somewhere nice and dry. When it finds a dry spot, it attaches itself firmly to the ground. About half its constituent amoebas get together to put up a long, thin stalk. The rest form a big round blob at the top of the stalk, known as a fruiting body. (I say 'long' and 'big', but that's relative to the size of a single amoeba. The stalk can be up to two centimetres high, and the fruiting body can be the size of a pea.) The amoebas in the fruiting body turn into spores, and blow away on the wind. The cycle starts anew.

You'd be justified in thinking that the slime mould amoeba must have some pretty clever genes – and no doubt it does. However, most of those genes just tell it how to be an amoeba. It is probable that the ones that help it form patterns mostly tell individual amoebas how to send out chemical signals, and how to respond to them. The actual patterns are not specified in the genes, not as such: instead they 'emerge' from the mathematical rules that these chemical signals, and the amoebas them-selves, obey. In fact, the entire life cycle of the slime mould may well owe at least as much to mathematics as to genetics☞.

I don't really want to talk about genes at all, but everyone is very aware these days that living organisms' genes contain DNA 'codes' that tell them what to do. So whenever we observe an organism doing something strange, our reflex assumption is that some gene, or a bunch of genes, is responsible. I don't think that's always so – in fact I think that even when it's true, most of the time it's an exaggeration. Many peculiarities in the structure and behaviour of organisms arise from a combination of genetics and mathematics, and some are almost entirely mathematical☛.

Whether I'm right or not, questions of natural pattern have given rise to one of the most elegant and beautiful ideas in the whole of mathematics: the concept of symmetry. I'm going to take an extensive look at the relation between symmetry and patterns. Eventually we'll pick up the tale of the slime mould and see how symmetry explains their spirals, and how the same kind of mathematics explains lots of related things like chemical patterns and the electrical activity of the heart. The deep message is that mathematics can reveal unexpected unities in nature, unities that go much further than mere coincidence or analogy.

THE SCIENCE OF PATTERNS

One of the best definitions of mathematics is 'the science of patterns'. Mathematics is how we detect, analyse, and classify regular patterns – be they numerical, geometric, or of some other kind.

But what is a pattern?

A pattern is a landmark in the magical maze. It's one of those things that you recognise when you see it, but it's not so easy to pin down the concept of a pattern once and for all with a neat, tidy, compact characterisation. In fact, the entire development of mathematics can be seen as a slow and erratic broadening of what we accept under the term 'pattern'.

To the ancient Babylonians, most mathematical patterns seem to have been numerical: they were, for instance, fascinated by the sequence of perfect squares 1, 4, 9, 16, 25, and so on. They even found a way to use tables of squares to multiply numbers together☛. To the mathematicians of ancient Greece, patterns were mainly geometric, and of a rather limited kind. Today's concept of pattern is considerably broader and more flexible.

To keep the discussion specific, I'm going to limit myself to a fairly restricted, but still very rich, class of patterns: those that possess symmetry. In fact this whole chapter can be thought of as an extended exercise on the topic 'what is symmetry?'

The word 'symmetry' is used in ordinary language in at least two different ways. One carries the connotation 'having a well-balanced form' – for example, we might praise the symmetry of a painting or a sculpture, but in very general terms, implying little more than a feeling that the work is rather well designed. The second meeting of 'symmetry' is much more specific, and not as closely related to aesthetics. We say that a form has symmetry if different parts of it repeat exactly the same shape or pattern. The commonest example of this usage occurs when we say that the human body is (approximately) symmetric, meaning that our left and right halves match, more or less exactly. This kind of symmetry is called *bilateral* – two-sided – and it is found in many living creatures. In the biological world, bilateral symmetry is never mathematically exact. For example, the human heart is usually on the left of the body, and there is no matching second heart on the right.

However, mathematics is *always* an idealisation of the real world, not an exact description. For example, a mathematical circle is perfectly round and infinitely thin. There are no exact circles in the real world – but that doesn't stop us saying that a car wheel is circular or that ripples on a pond form concentric circles when you toss a pebble into it. And it doesn't stop that description being useful and informative, either. The heart is very nearly symmetric, and it's situated *near* the centreline of the body. Even the exceptions to bilateral symmetry come quite close. So when we talk about symmetry, I'm not going to worry too much about slight imperfections in nature, compared with the mathematical ideal.

I advise you not to, either.

SYMMETRY

The general idea of symmetry has been around for thousands of years – the word goes back to ancient Greece: *sym* ('together') + *metron* ('measure'). However, the mathematical concept of symmetry was not made precise until the early 1800s☞. One reason why it took so long to achieve this goal is that symmetry is not a thing. It is a *transformation* – a way of moving things around. Indeed, symmetry is often a whole collection of transformations.

Let's look at an example of a symmetric shape, and try to pinpoint what features make it symmetric. I've chosen a five-pointed star, the kind you find on a wizard's hat or on top of a Christmas tree. A five-pointed star is symmetric – in fact it has 'fivefold symmetry', whatever that means.

What *does* it mean?

Figure 27 shows several different five-pointed stars. One of them is symmetric, in the mathematical sense; the rest are not. I don't think you will have a great deal of difficulty in working out which one it is; one star looks *much* more regular in shape than all the rest. Made your choice? It was the one on the right, wasn't it? And it looks more regular because all five points are the same shape and size, and they are spaced at equal angles to each other.

So we all know how to *recognise* symmetry, even though we don't yet

Figure 27   Some five-pointed stars. Which one is symmetric?

have a way to pin it down precisely. That's the next task.

Each star has five sharp 'points' extending from the centre. I want to talk about the points, but the word 'point' has a special meaning in mathematics – it means a very, very tiny dot, so tiny that it has position but no size (another idealisation). To avoid confusion, I'm going to talk about the 'arms' of the star. The key to the symmetry of the star is to think about comparing one arm with another.

How would we go about checking that the five arms are exactly the same shape and size?

Well, we could make a lot of measurements with a ruler (*metron* = 'measure') and see if they all match up (*sym* = 'together'). Alternatively, we could carry out the same task by tracing one arm on to a sheet of transparent plastic, and laying it on top of the other one to see if it fits. If it does, corresponding measurements are all the same; if it doesn't, some of them aren't.

In effect, what we're doing with the tracing is *moving* one of the arms and comparing it with the other one. And we're employing what's known as a rigid motion – a motion that keeps the shapes and sizes the same throughout. We have to do that, otherwise we can't use the movement to confirm or deny equalities of corresponding measurements. If we allowed the motion to stretch or compress the size – if we traced the arm on to a transparent elastic sheet, such as clingfilm, and pulled it about – we wouldn't be able to notice that in the left-hand star of Figure 27, the arms are different from one another.

There are several different kinds of rigid motion in the plane. If you slide a shape in some direction without letting it turn, so that its orientation doesn't change, you've performed a motion known as a translation (Figure 28(a)). This has nothing to do with rewriting sentences in a different language: it's a slightly archaic term, but it's the word that every mathematician in the world uses for such a motion. For simplicity, we could call it a slide.

Another important kind of rigid motion is a rotation – in plain

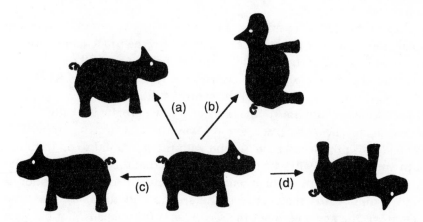

Figure 28 Symmetries in the plane: (a) translation, (b) rotation, (c) reflection, and (d) glide reflection

language, a turn (Figure 28(b)). Here you should imagine tracing the shape on to a sheet of transparent plastic, sticking a pin into the sheet somewhere, and then turning the entire sheet around that fixed pivot, through whatever angle you wish. The shape moves along with the sheet, and ends up in a new position *and* a new orientation. The place where you stuck the pin in is called the centre of rotation, and the angle you turned the sheet through is the angle of rotation. So we can say things like 'rotate a square through an angle of 45º about its top left-hand corner'.

The third kind of rigid motion is a reflection – or flip (Figure 28(c)). To reflect a shape, you can imagine placing a mirror (at right angles to the plane) in some position, and seeing where the shape *appears* to be when you look in the mirror. The position of the mirror is the *axis* of reflection. You get the same end result if you trace the shape on to a sheet of plastic, and flip the sheet over so that the other side is uppermost. (Incidentally, this 'motion' exemplifies a key feature of the way mathematicians think about transformations. What matters is where any given point ends up – not how it moves along the way. Reflecting in a mirror is not the same action as flipping a transparent sheet over – but both produce the same result.)

These are not the only possible types of rigid motion in the plane. A more subtle one, called a glide reflection (Figure 28(d)) reflects the shape in some axis and then translates it in a direction that is parallel to that axis, for example. But every rigid motion in the plane can be obtained by combining a translation, a rotation, and a reflection – perhaps omitting some of them.

Back to the stars. The simplest way to compare two arms of a star is to rotate the entire star about its centre, and see if the rotated arms overlap the original ones. If they do, then you can be sure that *all* the arms are the same shape. More subtly, you can also be sure that the angles between successive arms, measured from the centre of the star, are all the same.

Let's see why. Suppose you have a star, and you rotate it rigidly so that the arms all click one place round. Now the first arm has moved to coincide with the second, the second with the third, and so on until the fifth arm coincides with the first. This implies that the angle between the first and second arms – which remains unchanged because the motion is rigid – must equal that between the second and third. By the same argument, the angle between the second and third equals that between the third and fourth, and so on. We conclude that the angles between consecutive arms must *all* be equal.

We can even work out what those angles must be – *without* measuring them. Between them, the five equal angles amount to a full circle, 360º. So each angle must be one fifth of 360º, which is 72º.

Notice that we are now saying something a little stronger than 'the five arms have the same shape and are equally spaced'. Figure 29 shows a star with five arms, the same shape and equally spaced, but it doesn't look symmetrical – and it shouldn't. The use of rigid motions shows that when you rotate the entire 'symmetric' star through 72º about its centre, the result *looks exactly the same as the original*. If you weren't watching while the rotation was happening, you wouldn't notice the difference – unless you numbered the arms and moved the numbers with them. So the symmetry of the star is a property of the star *as a whole*, not just a resemblance between one arm and another. The star of Figure 29 doesn't fit properly when you rotate it, mainly because its arms, though they are the same shape and equally spaced, don't point in symmetrically related directions.

This is how mathematicians think about symmetry. A symmetry of

Figure 29 A star that is not symmetric, despite having five identical arms.

some chosen shape is a rigid motion that leaves the shape looking exactly the same as it was to begin with.

For example, instead of saying that the left and right sides of the human body have the same shape, we can observe that if we reflect the human body as a whole in an axis that runs vertically up the middle – say, by using a mirror – then the result is (to a good approximation) the same as the original.

Shapes can have more than one symmetry. In fact, the five-pointed star has several: rotations through the centre about angles of 72º, 144º, 216º, and 288º. For that matter, what about rotation through the full 360º? Or 432º, or …

Whoa back. One slight problem with the 'motion' description here is that it's a teensy bit too dynamic. As I just pointed out, we're not really interested in *how* the shape moves between its starting position and the finishing one. What matters when we want to establish the equality of corresponding measurements is where it ends up, and which arms in the original position correspond to which in the final position. From this point of view, a rotation through 360º is *the same* as one through 0º, and a rotation through 432º is *the same* as one through 72º. Don't think about the whole motion – just the beginning and the end, and what corresponds to what. Now, the 72º rotation is on our list already, so a rotation through 432º, say, is nothing new. But the 0º rotation isn't on our list, and it should be.

We can easily add it in, but we also need to sort out what it means. What kind of motion is a rotation through 0º?

A very simple one. *Don't move at all.*

Does not moving count as a 'rigid motion'? Rigid, yes; but linguistic purists might well object that it's not motion. However, it would be awkward to allow this objection to override the mathematical convenience of counting 'don't move' as a rigid motion. It's incredibly messy if you have to keep treating an angle of 0º as if it's a totally different manner of beast from angles of 1º, 2º, and so on. And, even worse, an angle of 360º as a totally different beast from 359º or 361º. Mathematicians dislike messy thinking, so they therefore sacrifice linguistic purity in favour of conceptual simplicity. In fact, linguistic purity is not a prominent feature of mathematicians, or of mathematics.

We've now found all the rotational symmetries of the star. Under any rotational symmetry, a given arm must move to one of five positions – the other arms, or itself – and once you know where one arm goes, the entire rotation is determined. Since we've already *found* five distinct rotational symmetries, we can be certain that we haven't missed any others. But have we found all the *symmetries* of the star?

No, we haven't.

There are five reflectional symmetries, too, whose axes run along the middle of each of the five arms. A little more thought shows that these are the only symmetries – because a given arm can move to only five positions, and once there, it must either be left where it is, or be flipped over. In particular, there are no translational symmetries, because if you move the entire star sideways then its centre has to move, in which case the star can't possibly fit on top of its original position.

THE KALEIDOSCOPE

Mirror symmetries, much like those of the star, provide the basis of a well-known toy, the kaleidoscope. This is a tube with an eyepiece to look through, and it looks a bit like a telescope, but it has no lens. If you look inside it, what you find is two bits of mirror joined together at an angle, with their common edge pointing down the middle of the tube. At the far end is either a random arrangement of coloured glass and plastic, or a place where you can insert a sheet of paper with a design on it. But if you look through the tube, what you see is a beautiful *symmetric* design.

Why? How can two mirrors – extending along only one side of their common edge to form a sort of wedge – produce a completely symmetric pattern? For that matter, if you look at the pattern you will find that, like the five-pointed star, it has rotational symmetries. Yet mirrors produce *reflections*. So where do those rotations come from?

When you look in a mirror, what you see is a reflection of the entire scene on your side of the mirror, except for bits that disappear from view because the mirror is of finite extent. What is actually happening, of course, is that light rays from real objects on your side of the mirror are hitting the mirror, bouncing off, and entering your eye from an unexpected direction – creating the illusion of objects 'behind' the mirror. It is this illusion that Lewis Carroll alluded to when he contrived to have his heroine climb through the looking-glass to visit the world on its other side.

With two mirrors, the light rays can bounce more than once. The effect is that 'behind' one mirror you appear to see not just the real objects on

Figure 30 Multiple refelections create the symmetries of the kaleidoscope.

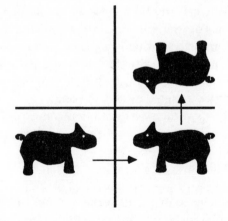

Figure 31 Two reflections combine to give a rotation.

your side of that mirror, but also reflections of objects in the other mirror. Figure 30 shows the effect that this has inside the kaleidoscope. Objects that lie in the angle between the two mirrors are reflected several times, and the combination of all those reflections creates the symmetric pattern.

Now, suppose that an object is reflected just twice – once in each mirror. What happens to it? Figure 31 shows that the combined effect of the two reflections is a rotation, whose centre is the place where the mirrors meet. Moreover, the angle of rotation is exactly *twice* the angle between the mirrors. So now we've learned several valuable lessons, among them that the result of combining two reflections, about mirror axes that meet at a common point, is a rotation. This result suggests that we take a closer look at what happens when several rigid motions are combined, by performing them in turn. In particular, what can we say if we combine two symmetries of some shape?

Well, the first motion leaves the shape looking exactly the same as when it started; and the second motion leaves the shape looking exactly the same as when it started. So clearly the combination of both motions *also* leaves the shape looking exactly the same as when it started. Conclusion: when you combine two symmetries of an object, you get a symmetry. This innocent-looking discovery has led to a huge amount of mathematics, a kind of 'algebra of symmetry', which makes such observations more systematic and forges from them a general toolkit for studying symmetry. This toolkit is called 'group theory'. We won't head off in that direction, but we shall take with us one useful consequence: whenever you've found a few symmetries of an object, you can often find more – or at least check that you haven't missed any 'obvious' ones – by seeing whether you get anything new by combining the symmetries you've already got. For example, once we know that a five-pointed star has a rotational symmetry through 72º, then we can combine that with itself to show that the star also has a rotational symmetry through $72º + 72º = 144º$. Combining this with yet another 72º rotation, we get a rotational symmetry through 216º, and so on.

When we find one symmetry, others often come 'for free'.

TIPOTS AND TOPITS

The kaleidoscope helps to explain another curious puzzle about mirrors. I'll lead up to it obliquely, by starting with a question which is often asked, but seldom correctly answered. It is this: 'Why does a mirror reverse right and left, but not top and bottom?'

It's a distinctly puzzling question, and unless you're careful, the more you think about it the more confused you'll become. Undeniably, if you stand in front of a mirror and wave your left hand, you see a figure in the mirror who appears to be waving their *right* hand. Whereas, if you stand in front of the mirror and wiggle your foot, what you *don't* see is somebody wiggling their head. So top and bottom stay the same, whereas left and right get swapped.

So why don't top and bottom swap if you turn the mirror on its side?

Hmmm.

Part of the difficulty is that we have got off on the wrong foot by thinking about a *symmetrical* object – ourselves. The problem with symmetrical objects is that several distinct rigid motions have the same effect on them – that is, after all, the definition of 'symmetric'. This opens up the possibility of confusing one such motion with a different one. In particular, there are *two* ways to make a bilaterally symmetrical object, such as the human body, coincide with itself after a rigid motion. One is to reflect it in a mirror plane, and that's the one under discussion. The other is to rotate it – in space – through 180º, and that's the one that's getting confused with the reflection when we try to think about the problem of left- and right-handedness. Our belief that a mirror interchanges left and right is a misconception, based on our own bilateral symmetry.

Our confusion about mirrors is made worse by a feature of human psychology that derives from a feature of the physical world. It is very easy to pick up an object and rotate it – or translate it. However, you cannot *pick up* an object and reflect it. If you could, then shoe factories could manufacture just right-footed shoes, and then move them around somehow to produce left-footed ones. For that matter, left- and right-footed shoes could mysteriously change their orientation whenever anyone moved them. Or, to be more accurate, we would no longer distinguish the left shoe from the right. Of course you can produce apparently reflected shoes by looking at them in a mirror – but, Alice's adventures notwithstanding, you can't reach into the mirror, pull out the reflected shoe, and put it on your other foot.

Translations and rotations, then, are the 'default' assumptions of our visual processing system. We think of other possibilities only if translations and rotations don't work. So when our brains see a mirror image of an object, their first reaction is to assume that it has been rotated. If we get a 'match' with a known object by doing this, we look no further. And that's what happens when we look at ourselves in a mirror: we get a match with a human figure. However, when we move our left hand, the matching figure moves *what would be its right hand if it were a rotated figure.*

However, it's a looking-glass figure in a looking-glass world, and *it* would think it was moving its left hand. The one on the same side as its heart, OK?

In order to avoid this kind of confusion – which is what causes the apparent difference between left–right and top–bottom reversal – we really ought to think about an asymmetrical object. My own favourite is a teapot shape, marked with eyes on one side (referred to from now on as its 'front'). In a distant galaxy there live two alien races. One is the Tipots, whose spouts are on their right and handles on their left. They are friendly, and they smile. The other is the Topits, whose spouts are on their left and handles on their right. They are distinctly unfriendly, and they frown.

What does a Tipot see when it looks at itself in a mirror? Not another Tipot. What it sees (Figure 32) is a Topit, with its spout and handle on the other side. There is no question of rotating, no confusion between the left hand and the right hand. No: the spout is on the spout side, the handle on the handle side. But there is something horribly wrong about the mirror image.

What?

Figure 32   A smiling Tipot looking in a mirror sees a smiling Topit.

Not spouts or handles, and not even eyes. No, it is the *smile* that is wrong. The image isn't a Topit at all, because it's smiling!

Why, the Tipots constantly ask themselves, does a mirror turn (what they know ought to be) a frown into a smile? Why does it swap top and bottom – but only on the mouth? It's a difficult philosophical question for the Tipots.

We may think *we've* got problems with mirrors, but theirs is even more baffling.

Of course, it is answered in the same way. Where we confuse reflection with a rotation (of a single bilaterally symmetric object), the Tipots confuse reflection with an interchange (of two symmetrically *related* objects). Both our species and theirs is falling into a psychological trap, interpreting the looking-glass world as if it were the normal one.

In fact, mirrors do not interchange left and right; neither do they interchange top and bottom. (Or smile and frown.) So the *question* was wrong.

What *do* they interchange?

Front and back.

Stand in front of a mirror, and see what bits of the image correspond to what bits of you – in the sense that a line drawn between those bits meets the mirror at right angles. Your head corresponds to your head. Your feet correspond to your feet. Waggle your left foot. The image may seem to waggle its right foot – but the foot that it waggles is still on the *left* side of the mirror, from your point of view. If you move your left hand, the image also moves the hand that is on your left. Ditto for the right hand.

However, if you are facing north, say, then the image seems to face south. Your front is to the north of your back, whereas the image's back is to the north of its front. Indeed, the effect of the reflection is to swap front and back along directions perpendicular to the mirror. Not by rotating the object, but by kind of squashing it flat and then opening it up the other way. Left, right, top, and bottom don't come into it at all – they *all* stay the same!

When you look in a mirror, you do not see yourself rotated. You see yourself reflected, and that's very different. You will immediately notice that it is different the first few times you see yourself on video. The face that stares out of the screen looks slightly wrong. It's not the familiar face that you see every morning in the bathroom mirror. Other people, however, fail to notice that it's wrong: to them the video image looks just like you.

The reason for this difference of opinion is that the human face is not perfectly symmetrical; moreover, we are so good at recognising differences between faces – the main way we identify other people – that we can distinguish the very tiny differences between our own face and its reflection. The rest of the world sees us as we are, but *we* normally see only our reflections. If we had eyes on stalks ... but we don't.

In order to see ourselves as others see us, we need a mirror that does not reflect our face, but rotates it. At first sight that seems impossible, but recall the discussion of kaleidoscopes. If two reflections are composed, the result is a rotation. Moreover, it is a rotation through twice the angle between the mirrors. Now, to look at a genuine copy of yourself in the bathroom mirror – one that is not reflected, but rotated – you want an image that faces the opposite way, a 180º rotation. Since 180º is twice 90º, we deduce that two mirrors at a right angle will produce the desired effect.

They do (Figure 33).

The main snag with this invention is that the line where the mirrors meet runs straight down the middle of your face when you look at yourself. Still, you can't have everything.

We are so accustomed to ordinary mirrors that using such a mirror is distinctly off-putting. I recently stayed with a friend in Singapore whose bathroom wall has two mirrors meeting at a corner. When I tried to shave, I kept moving the razor in the wrong direction. I tried to touch the

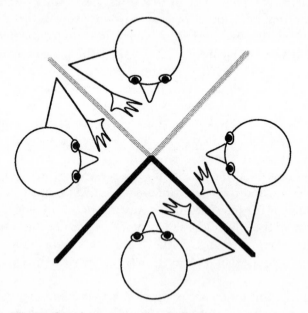

Figure 33   Mirrors that show you as you really are.

left side of my face, but my mirror image raised the wrong hand. Instead of raising the hand that 'corresponded' to my left hand – namely, the hand that lay on my left as I looked into the pair of mirrors – it raised the other one, the one to my right.

But for an image formed by a 180º rotation, this is the hand to the image's *left*.

So, ironically, *this* type of mirror actually does interchange left and right – *your* left and right. And it's not a comfortable feeling.

FRIEZES, WALLPAPER, AND ALL THAT ...

We've now encountered shapes with rotational symmetries, and with reflectional ones, but as yet none with translational symmetries. We could be forgiven for starting to wonder whether *any* shape can have translational symmetries. Suppose you have a shape that looks exactly the same after translation, say, through a distance of one metre in an easterly direction, to be specific. How can the shape fit on top of its old position if the whole thing is moved bodily sideways?

Let's think about that a bit more. If the shape fits exactly on top of itself when it is moved one metre to the east, then the same goes for the translated shape: it will *also* fit on top of its old position if it is moved one metre to the east. So the original shape must also fit on top of its original position if it is moved *two* metres to the east. The same argument holds if it is moved three, four, five ... any whole number of metres to the east. Indeed, running the argument backwards, the original shape must also fit on top of its original position if it is moved any whole number of metres to the *west*. All of which means that the 'shape' must extend infinitely far in both directions, repeating the same form indefinitely both to the east and to the west.

Nothing in the real world can do this, because real shapes are of finite extent. Indeed, our whole universe is of finite extent. But mathematics permits things to be infinite – for example, a line is infinitely long, the plane is infinitely wide, and there are infinitely many whole numbers. And with all this as a clue, it is easy to invent mathematical shapes with a translational symmetry. For example, here's one, using the letter 'L' as the basic repeating unit:

... LLLLLLLLLL ...

The dots ... indicate that the pattern continues for ever to both the left and right.

Interior decorators know about these types of pattern. Of course, they don't make them infinitely long, but they can make them as long as anybody wants – it depends on how big the room or passageway is. They call them 'friezes', and so shall we. Specifically, a *frieze* is a pattern obtained by repeating some basic unit over and over again in one direction. Here's a frieze with different symmetries:

$$\ldots \text{TTTTTTTTTT} \ldots$$

As well as translational symmetry, it has reflectional symmetry about any vertical axis that runs through the middle of one of the component T's. Any others? Yes: it also has reflectional symmetry about any vertical axis that runs through the middle of the space between two consecutive T's.

It's hard to think of any other frieze symmetries, until you start considering glide reflections. Then some more spring to mind:

$$\ldots \text{LГLГLГLГL} \ldots$$
$$\ldots \text{VΛVΛVΛVΛV} \ldots$$

And if we allow top–bottom flip symmetries, there's

$$\ldots \text{HHHHHHHHHH} \ldots$$

as well.

That's not all – but I'm running out of suitable letters, so I'll move on temporarily to another kind of pattern employed by interior decorators: wallpaper patterns. Over seventy years ago the mathematician George Polyá☞ proved that there are exactly nine basically different frieze patterns. He also went on to study the two-dimensional generalisation of the same question: *wallpaper patterns*. These also employ a repetitive basic unit, but now it repeats in two independent directions in the plane. Again, the pattern goes on for ever. And again, the important point is the symmetries of the pattern, not the particular shapes used as the basic unit. You can make wallpaper patterns by stacking friezes together, for instance:

```
       : : : : : : : : : :
... L L L L L L L L L L ...
... L L L L L L L L L L ...
... L L L L L L L L L L ...
... L L L L L L L L L L ...
       : : : : : : : : : :
```

You can also find some genuinely two-dimensional wallpaper patterns, which don't come from friezes; Polyá proved that there are exactly eight more of these. Figure 34 shows all seventeen wallpaper patterns. It's striking that, despite the enormous variety of patterns in wallpaper catalogues, they all fall into one or other of precisely seventeen symmetry types.

A novel variation on the wallpaper theme has recently been found by

Figure 34   Polyá's seventeen wallpaper patterns.

Marty Golubitsky and Ian Melbourne☞: they have classified the possible symmetries of architectural columns, showing that there are twenty-four possible kinds.

Symmetric patterns just like wallpaper patterns are especially prevalent in Islamic art, where it is normal to decorate mosques and other buildings with abstract mathematical designs. It is often said that this happened because representation of the human figure is forbidden in the Koran, but that's not the case. There is a body of figurative art too, notably miniatures. However, it is certainly true that Islamic art mostly avoids human figures, animals, and other shapes that are the staple of most Western art.

Islamic art, indeed, goes well beyond the rigid concept of mathematical symmetry. All seventeen possible wallpaper patterns are to be found in Islamic art☞, but there are many other patterns too. Many basic motifs in Islamic art are inspired by tilings – symmetric patterns in which a small range of shapes are fitted together to cover the plane. Western math-

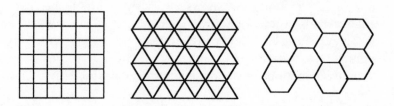

Figure 35 The three ways to tile the plane with identical regular polygons, arranged in the same manner at every vertex.

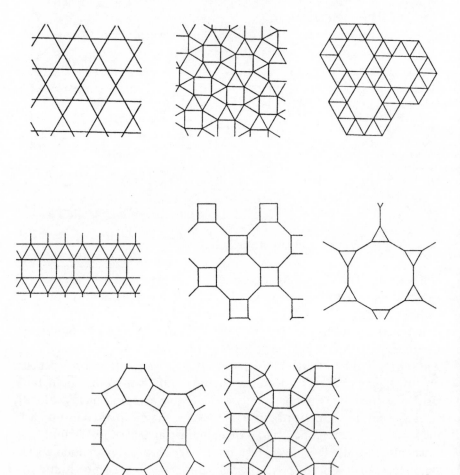

Figure 36 The eight ways to tile the plane with several regular polygons, arranged in the same manner at every vertex.

ematicians have classified the possible tilings that make use of regular polygons. (Regular polygons themselves have symmetry, like the five-pointed star: rotations through appropriate angles, and reflections in appropriate mirrors.) With one shape only, there are three: tilings by equilateral triangles, squares, or hexagons (Figure 35). The regular pentagon cannot tile the plane. If several regular polygons are used, there are more possibilities (Figure 36). But the Islamic artists found some beautiful patterns that look as if they are made from regular polygons – but aren't. Figure 37 shows my favourite example of such a pattern☛.

To see that the tiles can't all be perfectly regular, observe that at some corners apparently regular pentagons, hexagons, and heptagons meet. The angle at the corner of a pentagon is 108°, that for a hexagon is 120°, and that for a heptagon is $128^4/_7$°. These add up to $356^4/_7$°, but the total actually has to be 360°. Nevertheless, the artist has disguised the discrepancies so cleverly that the eye sees nothing amiss.

In recent years, Western mathematicians have also devised some

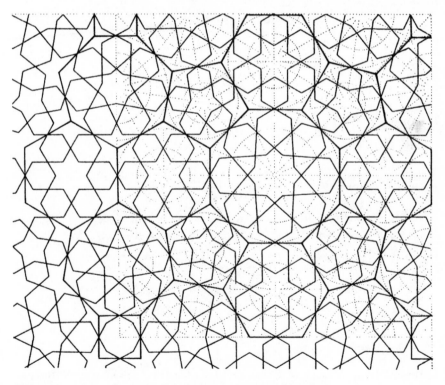

Figure 37   Islamic latticework based on an 'impossible' tiling.

astonishing almost-patterns, the most famous being Penrose tilings – named after the mathematical physicist Roger Penrose. Penrose tilings (Figure 38) never repeat exactly the same pattern – in fact they can't – and they have elements of fivefold symmetry.

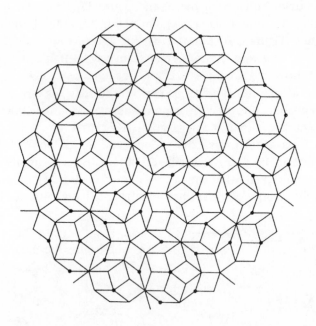

Figure 38   Penrose tilings never repeat, but cover the plane none the less.

GOLFBALL SYMMETRIES

Moving into three dimensions, we can seek the analogues of the regular polygons. They are the regular polyhedra, and there are five of them (Figure 39):

- tetrahedron
- cube
- octahedron
- dodecahedron
- icosahedron

The tetrahedron has four triangular faces and 24 distinct symmetries, the cube has six square faces and 48 symmetries, the octahedron has eight

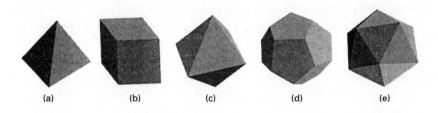

Figure 39   The five regular polyhedra: (a) tetrahedron, (b) cube, (c) octahedron, (d) dodecahedron, and (e) icosahedron.

triangular faces and 48 symmetries, the dodecahedron has twelve pentagonal faces and 120 symmetries, and the icosahedron has twenty triangular faces and 120 symmetries.

In every case the number of symmetries is equal to twice the number of faces, multiplied by the number of sides on a face. For example, for the dodecahedron we have $2 \times 12 \times 5 = 120$. This relationship is a consequence of the regular geometry of these solids. To see why it holds, imagine rotating a dodecahedron so that it occupies the same space that it started from. Choose a particular face. It can move to occupy any of the 12 face positions. Having chosen one such face, the dodecahedron can then be rotated to any one of five positions while preserving that face. That gives $12 \times 5 = 60$ rotational symmetries. But the dodecahedron can also be reflected before performing these rotations, and that doubles the final number of possibilities.

The octahedron and the cube have the same number of symmetries, as do the dodecahedron and the icosahedron. This is not a coincidence, but a consequence of what geometers call 'duality'. The mid-points of the faces of an octahedron form the vertices of a cube, and conversely; this implies that the octahedron and the cube have the same symmetry group. The same goes for the dodecahedron and the icosahedron. Soccer balls have the same symmetries – but differ in their actual shape – as an icosahedron. So do many viruses☞. But who would imagine that there is significant mathematics in the humble golf ball?

Not me, certainly, because until my attention was drawn to the fact by Tibor Tarnai☞, a Hungarian engineer, the possibility had never occurred to me. Golf balls have an interesting structure because they are not perfect mathematical spheres: they are made with patterns of dimples, which improve their aerodynamic performance by reducing drag. Manufacturers employ an amazing variety of dimple patterns, and nearly all of them are highly symmetric – which is where the mathematics comes in. The symmetries of dimple patterns are surprisingly varied: they range from the symmetries of a regular pentagon to those of a dodecahedron. Moreover, there are connections between the symmetries and the number of dimples, leading to 'magic' numbers of dimples that occur far more often than others, such as 252, 286, 332, 336, 360, 384, 392, 410, 416, 420, 422, 432, 440, 480, 492, 500 – all of which occur in commercial balls. One manufacturer made a ball with 1212 dimples, but not for actual play – just to show how clever they were at making golf balls.

Nowadays dimples are universal, but around the turn of the century golf balls were made with raised bumps, and were consequently known as 'bramble' balls. The dimples are tiny circular depressions, which must be distributed over the surface of the ball in a fairly even manner. If some regions of the surface have more dimples than others, then the ball will tend to swerve unevenly. So the problem of designing golf balls is closely related to that of distributing circular discs evenly over the surface of a sphere. However, there are practical considerations that must also be taken into account. In particular, the ball is moulded from two hemispheres, so there must always be at least one 'parting line' – a great circle on the sphere that does not meet any of the dimples. In order to ensure aerodynamic balance, most designs have several parting lines, arranged symmetrically.

The maximum number of symmetries for a golf ball is 120, but the number of dimples may be larger than this. Early golf balls used two distinct patterns. In the first, the dimples were arranged along 'lines of latitude' relative to an 'equatorial' parting line. Usually there were 392 dimples, and the golf ball had fivefold rotational symmetry about an axis running from its north pole to its south pole, but often the number was reduced to 384 because eight dimples were omitted for obscure reasons. The second pattern had the same symmetry type as the regular octahedron, a shape formed from eight equilateral triangles, like two square-based pyramids joined base to base. The classic octahedral golf ball has 336 dimples; it has three parting lines which meet each other at right angles, just like the Earth's equator and the meridians at 0° and 90°.

A more esoteric design possesses icosahedral symmetry with six parting lines. An icosahedron is formed from twenty equilateral triangles, arranged in fives around each vertex. The same arrangement is found in geodesic domes such as those used to house radar equipment, and in the protein units of many viruses. The adenovirus, for instance, has 252 protein units configured as an 'icosidodecahedron' (Figure 40), whose faces are either equilateral triangles arranged five around a vertex, or regular hexagons (which can be thought of as equilateral triangles

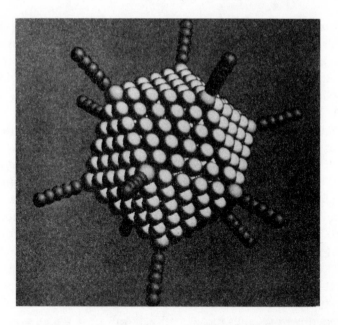

Figure 40   The icosahedrally symmetric adenovirus.

arranged *six* around a vertex). Uniroyal™ makes a golf ball with exactly this structure: unusually, its 252 dimples are five-sided pyramids.

SYMMETRICALLY SEEKING SLIME MOULD

Slime mould forms patterns – spiral ones. Can we understand slime mould from the point of view of symmetry?

We can, but it's a bit subtle, because the symmetry is not obvious in a single snapshot of the pattern. A fixed spiral is elegant in form, but it does not possess symmetry in the mathematical sense. However, a rotating spiral does possess a dynamic symmetry – as does any continuously rotating object. It is a symmetry that takes place in both space and time. Start with the spiral, let time pass. It rotates to a new position. Now rotate it backwards through the appropriate angle, and it looks exactly as it did to begin with. This is a *space–time* symmetry: translate time *and* rotate space. Indeed, the rotating spiral has lots of space–time symmetries, because every period of time corresponds to some spatial rotation. If waiting 10 seconds and rotating backwards by 10° leaves the spiral fixed, so does waiting 20 seconds and rotating backwards by 20°, or waiting 147 seconds and rotating backwards by 147°. In fact, if you wait x seconds and rotate backwards by $x°$, you won't notice any difference, whatever x may be.

Rotating spirals – and some other interesting patterns – can be observed in a beautiful chemistry experiment. Not the kind where you mix up a few chemicals and it all blows up in your face with a loud bang and a puff of purple–orange smoke: this is a gentler experiment, and instead of it doing something and then stopping, it carries on doing it for quite a long time. It is known as the Belousov–Zhabotinskii reaction, or BZ reaction for short – usually pronounced 'beezee' in the American style. B.P. Belousov was a Russian chemist who first devised the experiment in the 1950s, but he was too far ahead of his time and nobody believed that such a thing could happen. Since they knew it was impossible, they didn't bother to visit his laboratory and see. A.M. Zhabotinskii improved the choice of chemicals, and by then chemists had begun to realise that maybe chemical reactions that went through the same stages several times in a row might be possible after all, so people did visit his laboratory, and the profession allowed him to publish his work.

The experiment requires four chemicals. Any school laboratory will be able to get hold of them, and almost any university laboratory will already have them. Some are mildly poisonous, so it's not easy to buy them over the counter.

Mix up three of the chemicals in a beaker, following a simple recipe☞. The mixture is about the colour of a glass of lager. Leave the mix to clear, then add the fourth chemical, which is an indicator whose colour changes according to the acidity or alkalinity of the mixture. Pour the result into a Petri dish – a flat circular glass dish about ten centimetres across. The mixture goes pale blue, then orange–brown.

Sit back and wait.

After about five to ten minutes, suddenly you'll see one or more tiny blue dots. The dots slowly grow, and when they get big enough, tiny brown dots appear at their centres. This process continues, so that after ten or fifteen minutes you find a series of 'target patterns' of consecutive orange and blue rings, slowly expanding (Figure 41).

This effect is an example of spontaneous pattern formation, otherwise known as symmetry-breaking. Instead of the colour being uniform orange, a state which (in an ideal infinite dish) would be symmetric under *any* rigid motion, each ring system possesses only rotational symmetries about its centre, plus reflectional symmetries in axes that pass through its centre. (It helps not to be too literally minded here, and to think about just one system of rings in an infinitely large dish.)

Figure 41   The expanding rings of the BZ reaction.

Why don't the chemicals take up the fully symmetric uniform state? Because it is unstable. Any tiny lack of uniformity grows, and destroys the uniform pattern. And in the real world there are always tiny lacks of uniformity – dust motes, bubbles, even just a few molecules vibrating because of heat. (All molecules vibrate because of heat – or more accurately 'heat' is what you get when molecules vibrate – but it only takes a few of them to *trigger* instability.) The instability is not intuitively obvious, but it's what happens both in the real world and in mathematical models, and here we can take it as given.

If you now deliberately break up a target pattern, say by scraping your car keys through it – gently – then you can create spirals. The broken end of a blue ring, say, will start to curl up, wrapping itself round and round

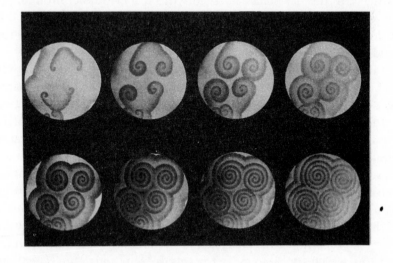

Figure 42   Spirals in the BZ reaction.

until an extensive spiral forms (Figure 42). The spiral keeps rotating, just like the slime mould. And, like the slime mould spiral, it has a whole family of space–time symmetries.

Where did those symmetries come from? From the even more extensive set of symmetries of the (idealised) uniform state in the infinite dish. The instability of that state caused certain symmetries to be eliminated, but *others persist*. For target patterns, some rotations and reflections persist. For spirals, what persists is the space–time symmetries 'let time pass and then rotate back'.

In a very curious sense, the patterns that we see in the spirals are evidence of other patterns that might have been – the unstable uniform state with its enormous amount of (totally boring) symmetry. They are 'caused' by something that doesn't actually happen.

TURING'S TIGER

The mathematics of patterns like those in the BZ reaction is the invention of Alan Turing. Turing was a British mathematician and logician, one of the pioneers of computer science. His ideas found concrete form in Colossus, a computer that was used during the Second World War to decode enemy messages at Bletchley Park in central England. Turing had many interests, and one of them was the markings of animals. When he was devising his mathematical theory, he would wander around with diagrams on sheets of paper, and show them to his colleagues. 'Doesn't that look like the markings on a cow?' he would ask excitedly – and they would nod, slowly, as if humouring the village idiot.

It's not yet clear whether Turing is going to have the last laugh – but his idea has certainly turned out to be important☛. What he did was to write down a system of mathematical equations, now known as reaction–diffusion equations. He imagined that the surface of an animal (more accurately, of an animal embryo, yet to grow into an adult) contained various chemicals. These could react with one another, and also slowly diffuse sideways. The BZ reaction is a case in point: the chemicals at any given location react to create changes in colour, and the chemicals can also spread slowly across the dish. This is why the colours move, but it's not the direct cause. That is, when we see a blue ring moving outwards, we are *not* observing diffusion as such. The motion we see is like waves on the ocean, which appear to move towards us even though the water is just going up and down. It is a wave of changing chemical concentrations.

But without diffusion, those waves would not be able to move, because the chemical reaction taking place at one point would have no effect on anything nearby.

Turing noticed that reaction–diffusion systems can spontaneously generate patterns. He found patterns of stripes, for example, and patterns of spots. Many animals have stripes – tigers, zebras, raccoons. Others have spots: leopards, giraffes. Turing suggested that chemical patterns in embryos could lay down a 'pre-pattern' which, when the animal developed, became the markings that we see. To be sure, you don't see target patterns or rotating spirals in animal coats – but different patterns can form under different conditions, so presumably animals could generate such patterns if nature found a good use for them. There is a Gary Larson 'Far Side' cartoon in which a deer consoles another deer who bears target pattern markings – an unfortunate incentive to hunters: 'Bummer of a birthmark, Hal.' Not that this has the slightest bearing on real animals, of course.

Biologists got as excited as Turing about this idea, but their experiments didn't agree completely with his theory. The stripes were spaced the wrong distance apart, or changed in the wrong way if the embryo grew at a different temperature. Or the biologists could watch how the individual cells of the embryo rearranged themselves to create the stripes, and it didn't look at all like a reaction–diffusion model. They decided that the similarities between the patterns formed by reaction–diffusion systems, and those seen on animals, must be coincidental.

Recently, however, Turing's ideas have been revived. Hans Meinhardt

Figure 43 Bednall's volute shell *Volutoconus bednalli* (left), and a computer simulation based on the Turing model (right).

has shown that nearly all the intricate markings on seashells can be explained by systems like Turing's (Figure 43). And Shigeru Kondo and Rihito Asai☞ have applied Turing models to the stripes on angelfish. They made the astonishing prediction that in a real angelfish those stripes would *move*, and that they would break up and re-form in characteristic ways (Figure 44).

They do.

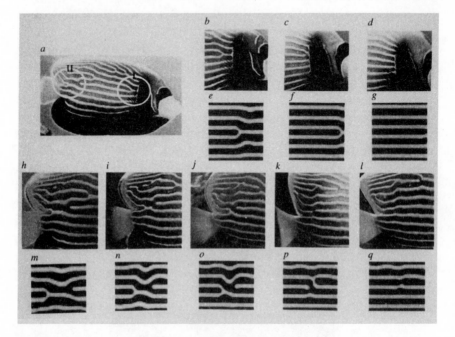

Figure 44 Disconnections and reconnections in an angelfish, and Turing simulations. (a) The angelfish *Pomacanthus imperator* with two regions circled; (b–d) changes in the head region, and (e–g) corresponding simulations; (h–l) changes in the tail region, and (m–q) corresponding simulations.

CELLULAR AUTOMATA

There is a competitor to Turing's theory, which turns up much the same patterns. The two ideas are fairly closely related, though this may not at first be apparent. The competitor makes use of a mathematical gadget called a cellular automaton, originally invented by the great Hungarian-born American John Von Neumann in a theoretical study of self-reproducing machines. For an image of a cellular automaton, think of a huge chessboard, each of whose cells can contain one of a fixed collection of markers – colours, say. At time zero the game is set up in some way. At each subsequent instant of time, the marker on each cell – its 'state' – is changed, depending on what the states of the neighbouring cells are, according to some fixed system of rules. For example, one rule might be 'a blue cell adjacent to two greens, one yellow and one pink should turn red'. Such a structure can contain well-defined 'objects' – arrangements of cells in a given shape and with given colours. A simple object might be a 4×4 square, all of whose cells are red. A more complicated one might involve a million cells, forming a vast, complicated shape, with particular colours in particular places.

By looking at cellular automata you can readily convince yourself that simple mathematical models can create not just spots and stripes, but some of the more unlikely markings found in the animal world. I'll use automata whose chessboard is filled up one row at a time, starting from some initial row – which I'll specify. I'll also specify rules for changing the colours as we move to successively lower rows.

For instance, assume just two colours, black and white. Suppose that we start with all cells black, and apply the following rule:

- The colour of any cell in the next row is the same as that of the cell immediately above it.

Then, of course, the next row will be all black, and so will the row after that. In fact, they'll *all* be black – and that's one way to get a totally uniform pattern (Figure 45(a)).

Next, suppose that we again start with all cells black, and apply a different rule:

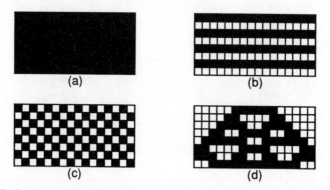

Figure 45 Creating patterns with cellular automata: (a) uniform, (b) stripes, (c) check 'spots', (d) chaotic triangles.

- The colour of any cell in the next row is the opposite to that of the cell immediately above it.

Then, of course, the next row will be all white. The row after that will be all black, and we'll get alternate rows of black cells and white ones. That's one way to get a striped pattern (Figure 45(b)).

For a third possibility, suppose that we apply the second rule above, but now start with cells that alternate black and white. Then the next row will also alternate black and white, but with white replaced by black and black by white. That yields a chequered pattern – in effect, spots (Figure 45(c)).

For a more exotic pattern, we start with a series of black cells, sandwiched between white ones. Now the rule is this:

- If the two cells immediately above a given cell, to its left and right, are the same, paint the new cell white. If they are different, paint it black.

The result is Figure 45(d). This is a very peculiar pattern indeed. It is so irregular that it scarcely deserves to be called a pattern at all. Yet lots of different cellular automata, with lots of rules, gives patterns of the same kind. And the pattern occurs in nature. For instance, the olive shell *Olivia porphyria* has exactly this kind of marking (Figure 46(a)). Seashells grow by adding material along their edge, so you can imagine that rules a bit like those for cellular automata might well apply. Surprisingly, Turing's reaction–diffusion systems can also produce these chaotic triangular patterns (Figure 46(b)). So there are *two* reasonable mathematical theories

(a)   (b)

Figure 46   (a) The olive shell *Olivia porphyria*; (b) a computer model based on a Turing reaction–diffusion system.

for the same strange patterns. The theories were found quite recently – 1956 for Turing's, the 1980s for cellular automata. The olive shell has been making its patterns for millions of years.

All of which suggests that nature knows a lot more about the mathematics of pattern formation than we do. But I reckon we're catching up.

JUNCTION FIVE

You turn the corner into the fifth passage of the magical maze.

You stop dead at the sight that greets you.

The floor, walls, and even some parts of the ceiling are covered in footprints. There are footprints of what look like mice, cats, and dogs; splay-footed, webbed prints of ducks; hoofmarks of cows, sheep, goats … Here are prints so large that they must have been made by elephants. The U-shaped hoofprints of shod horses are everywhere.

Shirley Combs would have had a field day here.

At intervals along the passage, lining the walls, are grandfather clocks – all identical. Their hands are still, their pendulums motionless. As you pass the first, it suddenly begins ticking. Did the vibrations of your feet set it in motion? Or something more sinister? Then the next clock begins to tick. Soon, as far as your eye can see, pendulums are swinging to and fro.

They are moving in strange patterns. Some are swinging in synchrony with each other.

Some aren't.

As you watch, the patterns keep changing. You can find no rhythm to them, yet you are convinced that their swings possess some mysterious kind of hidden order.

You begin to walk along the passage. The light dims. Fireflies dance near the walls, flashing at you in an eerie green light. At first the flashes are irregular, but after a time they begin to synchronise. Soon entire walls are flashing together, the light almost too bright to bear.

You move on down the passageway, towards a distant patch of what looks like daylight. Behind you, you leave a trail of glowing footprints where your feet have trodden.

Somewhere, a large audience claps in unison.

Now you hear clattering hooves, the sound of a galloping horse – or perhaps just a mobile coconut. The sounds grow to a crescendo – and then fade, as if the horse has passed you. Yet you see nothing. The sounds seem to come from beneath your feet.

You start to run.

The glowing footprints start to run, too.

Passage Five

THE PATTERN OF TINY FEET

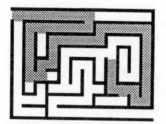

The serious scientist, I have often been told, does not waste his or her time on such trivial activities as writing articles for the newspapers, reviewing books, reading – or, perish the thought, writing – science fiction, or taking part in radio chat shows.

True. At least, *that* sort of serious scientist doesn't. The scientist who thinks that 'serious' means the same as 'solemn'. I guess I'm not that kind of scientist, but I don't consider myself particularly frivolous. I take science seriously. I just happen to think that we live in a world of many different dimensions, and that it is foolish to limit ourselves to just a few of them. In fact, I strongly believe that most scientists would benefit from expanding their horizons – not just in personal terms, but in their own specialist research.

I admit that it's very difficult to explain to someone whose main interest is X-ray stars that perhaps they ought to take a look at the latest research on the visual sense of lobsters: after all, there's enough serious stuff to learn about the latest in X-ray telescopes without getting side-tracked into zoology. As it happens – and I accept that I loaded my example – lobster vision is absolutely central to the latest in X-ray telescopes. Because X-rays are extremely energetic, it is impossible to focus them using lenses like those in normal optical telescopes. (That's why X-rays are so useful in medicine: they are much less affected by materials that have a strong effect on light. They can see through flesh.) However, X-rays can be reflected from flat surfaces if they graze them at a small angle. The lobster eye employs such an arrangement of mirrors, instead of a lens, in order to see. That arrangement turns out to be ideal for making an X-ray telescope, as the astronomer Roger Angel realised in 1978 when he read a paper on the lobster eye by Mike Land and Klaus Vogt. In 1996, a team of British scientists and an American company finally cracked the main technological problem in manufacturing an 'X-ray lens' based on the lobster-eye principle – fabricating the right kind of glass. The implications go further: the X-ray lens might lead to a new generation of computer chips, smaller than ever before, 'written' on a silicon wafer with a beam of X-rays.

Of course, examples of 'interdisciplinary research' with such spectacular consequences are fairly rare – but so are examples of *specialist* research with such spectacular consequences. Unless scientists immerse themselves in a rich environment of ideas, their vision will be as limited as that of a lobster with a lens.

Some years ago I became involved in a research area which, until that moment, I hadn't realised existed. The whole thing was triggered by a chance remark about an 'electronic cat' in a book review for the magazine *New Scientist*. To a 'serious scientist', in the limited sense described earlier, the whole thing would have seemed completely bizarre. To some of them it still does. For some reason known only to myself, I was squandering my already limited talents on a book review for a popular science magazine. Worse, I had wandered into an area totally removed from my normal field. I was working on pattern formation in dynamical systems – and there I was, reviewing a book on robotics, computer vision, and similar attempts to persuade machines to mimic some of the abilities of living creatures. About which I knew *nothing*.

From my point of view, however, this change of tack was entirely natural. It grew out of my main field of research, and – after some initial effort – it shed a new light on several old questions in that field. As we scramble along this, the fifth passageway of the magical maze, I want to tell you the story of the electronic cat, and its implications for mathematics.

AN ODD COINCIDENCE

The whole thing began when I noticed an odd coincidence.

I write a lot of book reviews. It's a great way to do three things at once: bone up on a new area, acquire a free book, and get paid for doing the above. It's also excellent practice in speed reading (the magazine or newspaper always wants the review by yesterday) and quick-fire journalism (ditto). *New Scientist* had signed me up to review a book called *Natural Computation*☛ – a collection of research articles on a variety of topics, whose common theme was to borrow a leaf from nature's book. The underlying philosophy was that, in order to design a system for computer vision, for instance, you should start by finding out some useful tricks from how animal brains process visual input; in order to design a legged robot, you should learn from how animals walk, and so on. Legged robots, by the way, are quite important: their applications include things like wandering round army firing ranges looking for unexploded mines, decommissioning nuclear reactors, and – one day soon – exploring the surface of Mars. (The current Mars robot has wheels – but wait.)

If that doesn't grab you, there are other reasons for understanding animal movements. For example, we can learn a lot about dinosaurs from the fossil tracks they have left in rocks that were once the muddy beds of puddles or shallow streams. We can reconstruct their behaviour – an especially poignant set of fossil tracks shows a number of tiny herbivore prints, on top of which are the heavy footprints of a pursuing carnivore. And we can reconstruct their movements. For example, *Triceratops* appears to have moved much like a modern rhinoceros. We can work out how fast they might have moved, and what kind of terrain they might have moved over.

Animal movements are fascinating, and important, for many, many reasons.

At the time the review book landed on my desk, I had been carrying out research with my friend and fellow-mathematician Marty Golubitsky on the natural dynamical patterns of 'coupled oscillator systems'. I'll explain what those are in a moment, but in concrete form they are like a lot of weights suspended by springs, so that they can move up and down, and hooked together by other springs, so that when one weight moves, they all do. We had discovered that if the weights are all the same, and the pattern of connections of the spring is a symmetrical one – say linking them together in a closed ring – then a particular catalogue of oscillation patterns always shows up, no matter what the precise details of the springs might be.

For instance, with two weights, there are two main patterns: both weights go up and down together; and when one goes up, the other goes down, and conversely. We had worked out a list of the general patterns for various arrangements of three weights, four weights, five, and so on.

I was thumbing through the review book, wondering what to write about it, when a paper on animal locomotion caught my eye. Animals move in a variety of patterns – for instance, a horse can walk, trot, canter, or gallop. These patterns are called *gaits*. An American zoologist called Milton Hildebrand had noticed that many of the commonest gaits are symmetrical, and his paper had been included in the review book. For instance, when a camel paces, both legs on the left move forward, then both legs on the right. It is just like two people inside a pantomime cow who are walking in step. A trotting horse, in contrast, is like two people in a pantomime cow who are walking *out* of step. When the front one moves her left leg, the rear one moves his right leg, and conversely. When an animal like a squirrel bounds, it is like two people in a pantomime cow who are doing bunny-hops. First the rear one hops forwards, both legs together, and then the front one does the same.

When the same thing happens in different places, or different times, it's often a sign of symmetry. Symmetries, after all, involve repeating patterns. The bounding squirrel, for instance, is left–right symmetric at every instant of time. There are more subtle symmetries, too: the front of the squirrel does the same thing as the back, but after a small time-lag. I don't recommend doing this with a real squirrel, but conceptually there is a second symmetry: swap front and back legs and wait for a moment. Hildebrand's point was that the main animal gaits all have symmetries of this general type.

However, he didn't ask *why* they were symmetrical. He just observed that they were.

Symmetry is an obsession of mine, especially symmetry in dynamics – in things that move. The symmetry of a moving object influences its motions in innumerable ways. I've spent a lot of my research life finding out how symmetry affects dynamics, and my mind is always on the alert for new examples. I was very struck by the gait patterns, because they were exactly the kind of behaviour that Golubitsky and I had found in symmetrical systems of four oscillators. Could it be that in some sense each leg of the animal either acted like an oscillator, or was controlled by an oscillator? Were the gait patterns clues to how those oscillators were coupled together? I knew nothing about gaits, so I couldn't make progress on those questions – but when I wrote my review I pointed out the coincidence. I also mentioned that our predictions about coupled oscillator networks had been tested using electronic circuits, and asked 'does anyone want to fund an electronic cat?'

The day after the review appeared, I got a telephone call from a young American physiologist called Jim Collins, who was studying in Oxford for a year. 'I can't fund an electronic cat, but I know people who can. I'd like to come and visit.' Thus began a collaboration that continues to this day.

FRIEZING TIME

Dynamics involves a crucial element: *time*. Symmetry is about shapes, which are things in *space*. How can symmetries apply to time?

We've just navigated our way along Passage Four, which was about patterns in space. It turns out that we can apply the same ideas to patterns in time, too. We can even put both together, and look at patterns in space and time combined.

Turning space into time is a typical mathematician's trick, a kind of 'technology transfer'. However, the 'technology' involved is conceptual technology – ideas – rather than gadgetry. No mathematician would ever waste a good idea by restricting it to just one incarnation. Many physically different systems possess an underlying mathematical identity: they obey rules which, in the abstract, are exactly the same – and must therefore lead to corresponding conclusions. The numerical concept 'two', for instance, applies to virtually anything: two chairs, two sheep, two dollars, two kilos of butter. This is exactly what gives mathematics its magical feel: we really don't know *why* the same concept can be so effective in so many different settings. There are all sorts of theories, ranging from the way we extract mathematical patterns from the world around us, to the possibility that we are merely selecting patterns that fit our own prejudices – but none are truly conclusive. Ultimately, I suspect, mathematics is an effective way to understand the universe because the universe happens to be like that.

From a physical point of view, time is very different from space – and this difference is often built into mathematical systems that model the physical world. Sometimes, however, we can exploit the very general nature of mathematics, and from this point of view a spatial distance and an interval of time are simply the values of certain numerical quantities, or *variables*. The difference between space and time then becomes a matter of interpretation – the same underlying mathematics can have several different meanings. Even by talking of 'patterns in time', I am taking that step: a pattern, after all, is normally thought of as a static design, a fixed form. 'Pattern in time' is then a very curious phrase, and an apparently meaningless one.

With just a tiny bit of lateral thinking, however, its meaning becomes crystal clear.

We've just been looking at frieze patterns – shapes that repeat indefinitely along a line, at regular intervals. A line normally represents space, but if we reinterpret that line as representing the passage of time, then presumably we are looking at 'shapes that repeat indefinitely along a *temporal* line at regular intervals'. Well, nearly. 'Shape' is the wrong word for something that exists in time. What word *should* we use? Let's reason it out by analogy – one of the sharpest weapons in the mathematical armoury. A shape is a (more or less) complicated collection of points in space; the temporal analogue is a complicated collection of points in time. A point in time is usually called an 'event'. So instead of 'shape' we have a 'series of events'. That is, the temporal pattern corresponding to the simplest frieze is a series of events that repeats indefinitely at regular intervals of time.

Can you think of any events like that?

Here are a few examples:

- Spring, summer, autumn, winter.
- Sunday, Monday, Tuesday, Wednesday, Thursday, Friday, Saturday.
- A beating heart.
- A swinging clock pendulum.
- A cork bobbing up and down in the sea as the waves pass.

Being physical things, none of these has *perfect* mathematical regularity, and of course they don't actually go on for ever. But, to a good approximation, we can *model* them by perfect, indefinitely repeating mathematical patterns – just as we model a finite frieze by an infinitely long mathematical pattern.

At any rate, the temporal analogue of the simplest frieze is a series of events that repeats over and over again. Such events are said to be *periodic*; the time required for them to repeat is the *period*. The first series of events above, for instance, has a period of one year.

It is useful to have a name for the abstract concept that all these examples have in common. 'Time frieze' would do, but for various reasons it's not what people use. Any mathematical system that varies periodically in time is called an *oscillator*. 'Oscillate' is a fancy word for 'wobble'. Oscillators are things that wobble, things that go up and down, things that go from side to side, things that rotate.

The world is full of oscillators. Physical oscillators range from galaxies, which revolve in a stately dance that takes billions of years to complete one turn, to subatomic particles, whose period of oscillation is a tiny, tiny fraction of a second. Biological oscillators are especially prevalent – the heartbeat, breathing, the sleep–wake cycle, the digestive cycle, wriggling, walking, flying, swimming ... Many of these biological cycles involve motion. When you walk, for example, you employ two oscillators – your legs. In fact you use other oscillators too, for example your arms. Each individual muscle involved in walking undergoes an oscillatory cycle. Indeed, each individual muscle *fibre* undergoes an oscillatory cycle. And the motion of your muscles is controlled by yet more oscillators, your nerve cells ...

But before we can understand whole systems of oscillators, we have to understand the simplest individual oscillators.

WHY DO OSCILLATORS OSCILLATE?

There are two physical examples of an oscillator which are especially suitable for carrying out thought experiments: the pendulum and the spring. A pendulum is a long rod, pivoted at one end, with a weight at the other, generally known as a bob. Usually the pendulum is confined to a vertical plane. Typically, a pendulum swings to and fro periodically – and we shall shortly see why. A spring is a long coil of wire, usually with a weight on the end; typically the weight bobs up and down periodically. Once I've explained the periodicity of the pendulum, you'll be able to give a similar explanation for the periodicity of the spring.

Why does a pendulum oscillate? We can deduce the periodicity from one general physical principle, without solving any equations or doing any calculations. All we need are two powerful ideas: energy and symmetry.

Energy is a measure of a system's capacity to perform work, and work is what gets done when a force moves. The bob at the end of a pendulum exerts a force – if you don't believe me, try picking up a suitcase. Strictly speaking, it is the Earth's gravity that exerts the force, and it *acts* on the bob. But, at any rate, lifting the bob requires work or, equivalently, uses energy. The kind of energy that you exert when you lift a weight against gravity is called *potential energy*. The amount of potential energy is proportional to the height and to the weight – a fancy way to say that moving the same weight to twice the height, or twice the weight to the same height, doubles the potential energy. Likewise for three times, or whatever.

Moving objects have another type of energy: *kinetic energy*. Like potential energy, kinetic energy is proportional to mass. It is also proportional to the *square* of the velocity – an object moving twice as fast has four times the kinetic energy; one moving three times as fast has nine times the kinetic energy, and so on.

The next thing you need to know is that – in the absence of friction – the total energy, potential plus kinetic, stays the same throughout the motion. (Friction reduces the total energy by converting some of it into a third form of energy, heat☞.) I'll now argue that this principle implies that the pendulum keeps repeating the same motions, over and over again. Figure 47 tells the whole story, but here it comes in words, too.

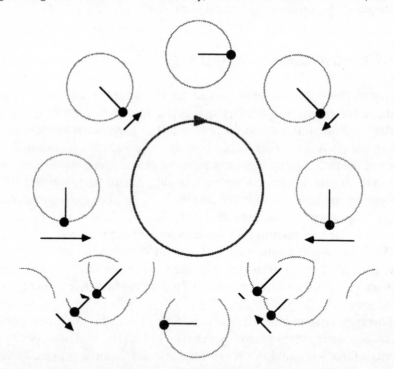

Figure 47 Why the pendulum oscillates. Clockwise from top: • Potential energy positive, kinetic energy zero. • Height drops, speed increases, energy stays the same. • At the lowest point, the speed is highest. • As the bob rises, its speed must drop. • When it comes to a momentary halt, it returns to exactly its initial height because the total energy doesn't change. ••• Now the sequence flips left–right and repeats, and the cycle starts anew.

A pendulum swings in a circle, so there is a lowest point that it can possibly reach – hanging vertically downwards. Any change in potential energy must be compensated by an equal and opposite change in kinetic energy: if the bob gets higher, it has to move more slowly; if it gets lower, it has to move faster. Imagine grabbing the pendulum by its bob and raising it to some chosen height. Hold it steady. Your hand did some work, and increased the bob's potential energy. The kinetic energy is zero because the pendulum is not moving. Now let go, and from now on let the pendulum move as it wishes. Gravity pulls the bob downwards, so the potential energy decreases. Kinetic energy must increase to compensate, so the weight starts to fall. At some stage the bob must reach the lowest point of the circle: here its potential energy is zero, so all of the energy has become kinetic. Not only is it moving at this point – it is moving as fast as it can possibly go.

Now the continuing motion along the circle takes it past the lowest point, and it starts to swing upwards again. The potential energy increases, so the kinetic energy must decrease: the pendulum slows down. How high does it go? It can't rise *above* its original height, otherwise it would have more potential energy than the total energy that it started with. So it has to rise to some level and pause, with instantaneous velocity zero. At this topmost limit, all of its energy is potential – so it equals the energy at the start, which is also all potential. The height it reaches must therefore be equal to the height from which it was originally released.

The same argument, repeated, shows that after a further period of time the bob returns to its exact original position, with its original velocity of zero. Having got back to where it started from, it is now obliged to repeat the same motions over and over again, for ever.

This kind of argument applies to any *deterministic* system – one whose future is always determined by its present state. In the absence of random effects, which can change the future motion in unpredictable ways, dynamical systems are deterministic. So once they return to their original state, they have no choice but to repeat the motions in between, over and over again.

This is why oscillators oscillate: their dynamics impels them to return to their original state.

We can say a little more: the motion of the pendulum has a certain kind of symmetry. The bob's motion during the second half of its travels is a left–right mirror image reversal of the motion during the first half. It therefore takes the same time as the first half – which must be half the overall period. So the deep cause of the pendulum's periodicity is the fact that a circle is left–right symmetric. This symmetry is showing up as a symmetry of the periodic motion – but it is a *different* symmetry. The symmetry of the circle is just a spatial reflection; that of the motion is a mixture of a reflection (space) and a delay of half the overall period (time). It is an example of a *space–time* symmetry. And it is these symmetries that govern the underlying patterns of such things as animal motion.

HUYGENS' CLOCKS

Before we bring on the elephants, however, we must introduce one final idea: that of coupling oscillators together. The first occasion that anyone noticed anything interesting about coupled oscillators was in February 1665, when the great Dutch physicist Christiaan Huygens, inventor of the pendulum clock, was confined to his room by a minor illness. One day, with nothing better to do, he was staring aimlessly at two clocks he had recently built, hanging side by side on the same shelf.

Suddenly he noticed something odd. The two pendulums were swinging in perfect synchrony.

Huygens was a good clockmaker, and he knew that his clocks didn't keep such good time that two of them would stay in synch. Yet these two did. It is a measure of Huygens' genius that he recognised the existence of a baffling problem – and that he eventually found an answer.

How perfect was the synchrony? Huygens watched his two clocks for hours, but they never broke step. Then he tried disturbing them – and within half an hour they had regained synchrony. Their synchronous motion was stable, not destroyed by a disturbance. Something must be stabilising it.

What?

If the clocks weren't 'communicating' in some manner, neither would 'know' if the other was falling out of step. Huygens realised that the clocks must somehow be influencing each other, perhaps through tiny air movements or imperceptible vibrations in their common support. Sure enough, when he moved them to opposite sides of the room, the clocks gradually fell out of step, until one was losing five seconds a day relative to the other.

Huygens' discovery gave rise to an entire sub-branch of mathematics: the theory of coupled oscillators. Coupled oscillators can be found throughout the natural world, but they are especially conspicuous in living things: pacemaker cells in the heart; insulin-secreting cells in the pancreas; and neural networks that control rhythmic behaviours like running, chewing, breathing ... even the waves of muscular activity that push food through the digestive system.

Indeed, the oscillators that are coupled may even be in different organisms. If a large audience of people are told to clap their hands rhythmically, they often synchronise the claps. If two people walk side by side, they often synchronise their steps. Alternatively, they may anti-synchronise them, so that when one of them moves her left leg, the other moves his right leg. This is a kind of 'reflected synchrony' – more subtle than perfect synchrony, and a close mathematical relative.

A single oscillator behaves in a very simple manner, repeating the same motions over and over again. When two or more oscillators are coupled, however, the range of possible behaviours becomes much more complex. Much of the complexity can be captured using the concept of 'relative phase': this is the difference in timing between the motion of two oscillators, measured as a fraction of the periodic cycle. For example, when two people walk in step, they carry out the same motions *at the same time*, so the relative phase is zero. When they walk in perfect anti-synchrony, one of them moves their left leg half-way between the times when the other does, and the relative phase is 1/2.

When an elephant walks, it moves first its left rear leg, then the left front, then the right rear, then the right front (Figure 48). Those movements are equally spaced, like four beats to the bar in marching music (Figure 49). Many other gaits, in many other animals, display similar patterns. The relative phases for the main quadruped gaits are shown in Figure 50☞.

SYNCHRONOUSLY FLASHING FIREFLIES

We'll say that two oscillators are *identical* if we can sensibly model them by the same system. Two pendulums with the same length of rod and the same mass for the bob are identical. Left to their own devices, they will behave in exactly the same way under the same conditions. The same goes for identical springs. In this chapter, all the oscillators in any coupled system will be considered to be identical. This, as usual, is an idealisation. An elephant's front leg is not quite the same as its rear leg. But their motion is very similar, so we'll agree to *pretend* that they are identical, and see where that leads.

Figure 48   The walk of an elephant.

Figure 49   The elephant march, drawn on an unorthodox four-line stave.

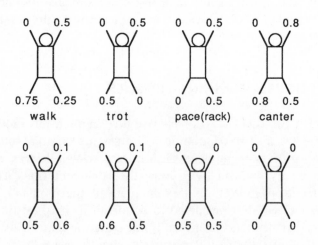

Figure 50   Relative phases for the main quadrupled gaits.

A general principle, first proposed by the physicist Pierre Curie, leads to the conclusion that the natural state for a system of identical coupled oscillators is synchrony – zero relative phase. Curie's principle says that symmetrical causes should have symmetrical effects. Coupled identical oscillators are a system with complete symmetry: *any* two oscillators can be interchanged without any visible effect on the system. According to Curie's principle, their motions should also be interchangeable – that is, they should all move in exactly the same manner. This is synchrony.

Perhaps the most spectacular success of Curie's principle among biological oscillators can be seen along the tidal rivers of Malaysia, Thailand, and New Guinea, where thousands of male fireflies gather in trees at night and flash on and off in unison in an attempt to attract the females that cruise overhead. When the males arrive at dusk, their flickerings are uncoordinated. As the night deepens, pockets of synchrony begin to emerge and grow. Eventually whole trees – even whole riverbanks – pulsate in a silent, hypnotic concert that continues for hours.

Curiously, even though the fireflies' display demonstrates coupled oscillation on a grand scale, the details of their behaviour were thought to be mathematically indecipherable. Fireflies are a 'pulse-coupled' oscillator system: they interact only when one sees the sudden flash of another and shifts its rhythm accordingly. Pulse-coupling is common in biology – crickets get excited when they hear another cricket chirping, and chirp back; neurons send signals to other neurons using electrical spikes – but it is hard to include pulse-coupling in mathematical models. Recently, however, Steve Strogatz and Renato Mirollo created an idealised mathematical model of fireflies and other pulse-coupled oscillator systems, and managed to prove that, under certain circumstances, pulse-coupled oscillators started at different times will *always* synchronise.

Their work was inspired by a 1975 investigation by Charlie Peskin on a model of the heart's natural pacemaker – a cluster of about ten thousand nerve cells known as the sinoatrial node. Peskin hoped to discover how these cells synchronise their individual electrical rhythms to generate a normal heartbeat, and he modelled the pacemaker as a large number of identical oscillators, each coupled equally strongly to all the others. A distinctive feature of Peskin's model was its physiologically plausible form of pulse-coupling. Each oscillator affects the others *only* when it fires. It kicks their voltage up by a fixed amount. And if any cells' voltage exceeds the threshold, then it fires immediately.

Peskin convinced himself that his model system will always become synchronised. He couldn't prove it, though: there were no established mathematical procedures for handling arbitrarily large systems of oscillators, and no effective methods for dealing with pulse-coupling. He therefore opened with a standard mathematician's gambit: focus on the simplest possible case, which here is two identical oscillators. He was hoping to solve that case in a manner that might let him 'let $2 = n$', but events didn't pan out so nicely. He restricted the problem further by allowing only very tiny kicks, and now the problem became manageable. For this very special case, he proved his synchrony conjecture.

The extension to an arbitrary number of oscillators, and any size of kick, eluded proof for about fifteen years. In 1989, Strogatz wrote a computer program to simulate Peskin's model for any number of identical oscillators. The results were definitive: the oscillators *always* ended up firing in unison. Strogatz and Mirollo reviewed Peskin's proof of the two-oscillator case, and noticed that it could be clarified by using a more abstract model for the individual oscillators. The key feature of the model turned out to be the slowing upward curve of voltage as it rose towards the firing threshold. Other characteristics were unimportant. From this more abstract viewpoint, Mirollo and Strogatz worked out how to play the 'let $2 = n$' game. And they proved that the system synchronises for almost all initial conditions, and for any number of oscillators. The fireflies had been happily synchronising their flashes all along – but now the mathematicians had a fair idea of what was causing the synchrony, at least in one reasonable but simplified model. For Peskin's model, as for the fireflies themselves, Curie's principle was vindicated.

SYMMETRY-BREAKING

Although Curie's principle helps to explain how coupled oscillators synchronise, it is not infallible. Symmetric systems don't *have* to synchronise – if they did, humans wouldn't be able to walk. Our legs would have to synchronise, with both moving forwards at the same time. This is not a recipe for elegant locomotion.

We can walk, despite our bilateral symmetry, because in some circumstances Curie's principle fails. Symmetric systems can undergo 'symmetry-breaking', in which a single symmetric state is replaced by several less symmetric states that together embody the original symmetry. Coupled oscillators are a rich source of symmetry-breaking: although some arrangements of coupled oscillators synchronise, many others do not.

The reason for this is that synchrony is just the simplest case of a general effect known as phase-locking – many oscillators tracing out the same pattern, but not necessarily in step. I mentioned earlier that when two identical oscillators are coupled, there are two main possibilities: synchrony, a phase difference of zero, and antisynchrony, a phase difference of half a period. Synchronous behaviour preserves symmetry; antisynchrony breaks it.

A mechanical analogy may help explain why these two motions are natural. Other types of motion are possible too, depending on the precise form of the oscillators – but these two motions arise for reasons of symmetry, and are therefore very general. Let's build on our experience with the pendulum, and see what happens when two identical pendulums are coupled together by a spring, as in Figure 51. Each pendulum is pivoted from a horizontal rod. A spring runs along the rod, coiling round it, and at each end it is fixed to a pendulum. When both pendulums hang vertically downwards, the spring is in its 'natural' state and exerts no forces. But as the pendulums swing, the coils of the spring tighten or loosen, creating new forces that tend to oppose the pendulums' motion.

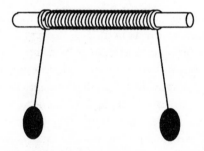

Figure 51   Coupling two pendulums by a spring.

To avoid confusion, let me emphasise that the spring does not lengthen or shorten, the usual motion of a spring: here it *twists*. And to keep things simple, I'll assume that the force it exerts is proportional to the amount of twist, and opposes the twist. No twist, no force. Twist the spring clockwise, opening its coils up – it wants to close again. Twist it the same amount the other way, tightening the coils – now it wants to open up, with the same force.

Synchrony first. Grasp both pendulums and rotate them away from the vertical line so that they hang at the same angle. Because their relative positions are unchanged, the spring remains in its natural state, neither stretched nor compressed, so it exerts no forces. If you now let go of both pendulums simultaneously, they are both starting from an identical height, so they begin to move in synchrony with each other – as if the spring were not there at all. Moreover, because they stay in synchrony, the spring *remains* in its natural state, so it has no effect throughout the entire motion. In other words, the two pendulums stay in synchrony, and oscillate in phase with each other.

At first sight this is a rather boring motion, because the spring doesn't actually play a role. Why mention it at all, then? However, a little extra thought shows that the spring coupling helps to stabilise the synchronous motion. Suppose that there were no connecting spring at all, and suppose that the two pendulums are not quite identical. When left to swing independently, any loss of synchrony will go uncorrected, and after a while their motions will bear no special relationship to each other. Now put the spring back. As soon as the pendulums start to go out of synch, the spring will be twisted away from its natural state. As a result, it will produce twisting forces that try to restore it to the natural state. But the natural state occurs only when both pendulums are at the same angle to the rod. In short, the presence of the spring will tend to restore synchrony.

Much the same happened with Huygens' clocks, but using vibrations transmitted along the common shelf.

This argument is fine if the differences between the angles of the two pendulums is small, but it breaks down if the difference is large. It is here that we find the antisynchronous oscillation. Imagine twisting the two pendulums through the same angle but in *opposite* directions – one clockwise, one anticlockwise. Their two angles are symmetrically related: each is a reflection of the other about the centreline. Now the spring exerts a force on them both. By symmetry, those forces are equal in size but opposite in direction. So when you let go, the two pendulums swing back towards the centreline – moving, as it happens, rather faster than they would have done if there were no spring. No matter: the important feature is that they *remain* symmetrically placed. So when one of them swings through the centreline going clockwise, the other must simultaneously swing through the centreline going anticlockwise. From that stage onwards, they both climb back to the same height as before, but on the other side of the centreline. Again, the total energy does not change during the motion, and this tells us that they rise to exactly the same height as they started from. And, throughout the entire motion, one pendulum *always* sits at an angle that is the mirror image of the other. So they remain exactly out of synchrony – that is, half a period apart.

And here we already see an intriguing connection with gaits. When a kangaroo hops across the Australian outback, its powerful hind legs oscillate periodically, and both hit the ground at the same instant: synchrony. When a human hunter runs after the kangaroo, however, his legs hit the ground alternately: antisynchrony.

PATTERNS IN NETWORKS

If the network has more than two oscillators, the range of possibilities increases. In 1985 Golubitsky and I developed a mathematical classification of the oscillation patterns of symmetrical networks of coupled oscillators. We found that systems of *identical* coupled oscillators typically exhibit phase-locking patterns. That is, the oscillators all do the same thing, but not necessarily at the same time. We've already seen this with two oscillators: they are either synchronised (and do the same thing at the same time) or they are antisynchronised (and do the same thing but half a period apart). In more complicated networks, the time difference is usually some fraction of the period. In some 'exotic' patterns one or more of the oscillators may actually be forced to behave differently from its fellows, but these patterns are not as common as the phase-locked ones.

We also found that for a given network several different patterns of oscillation are possible. Which one occurs depends on the 'parameters' of the system – adjustable constants such as coupling strengths. In any particular network, all the details of these phase-locked patterns can be worked out using the methods that we introduced.

For example, if four identical oscillators are coupled in a ring, with *directed* coupling (Figure 52(a)) – meaning that each influences the next one but not the previous one – then they can be phase-locked in four basic patterns:

Figure 52 (a) Directed coupling of four oscillators (circles) in a ring – schematic; (b) the same oscillators arranged to mimic the phase relationships of the elephant's walk.

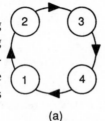
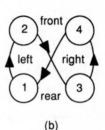

(a)   (b)

- All oscillators move synchronously.
- Successive oscillators around the ring move 'in turn' so that their phases differ by 0, 1/4, 1/2, 3/4.
- Successive oscillators around the ring move 'in turn' so that their phases differ by 0, 3/4, 1/2, 1/4.
- The oscillators form two groups. The first and third move in synchrony with each other, and so do the second and fourth, but the two groups are half a period out of phase. So the phases, read off round the ring, are 0, 1/2, 0, 1/2.

Collins and I noticed that there are striking analogies between these patterns of phase-locking and the symmetries of animal gaits such as the trot, pace, and gallop. In fact, quadruped gaits closely resemble the natural patterns of four-oscillator systems.

When a rabbit bounds, for example, it moves its front legs together, then its back legs. There is a phase difference of zero between the two front legs, and 1/2 between front and back legs. The pace of a giraffe is similar, but now the front and rear legs on each side are the ones that move together. When a horse trots, the phase-locking occurs in diagonal fashion. What about perfect synchrony? Young gazelles perform the 'pronk', a four-legged leap in which all four feet leave the ground together.

We could also relate the walking gait of a quadruped to four-oscillator networks. An ambling elephant lifts each foot in turn, with phase differences of 1/4 at each stage. This is just like the second and third patterns listed above, but the sequence in the elephant goes back left, front left, back right, front right, so the ordering of the oscillators has to be changed to move in a kind of 'figure eight' (Figure 52(b)). We had more trouble with three less symmetric gaits – the rotary gallop, the transverse gallop, and the canter. But in computer simulations of coupled oscillator networks, we could find similar patterns to those, too. Only we didn't really know why.

Collins and I discovered that the same methods seemed to work for insects, which have six legs. One of the stablest insect gaits is the tripod gait of a cockroach. Here the insect lifts three legs into the air, leaving a tripod of others on the ground; then it puts them down and lifts the other three. The pattern is like this:

- Phase 0: back left, middle right, front left.
- Phase 1/2: back right, middle left, front right.

We discovered that exactly this pattern arises in a ring of six oscillators. So do a number of the other observed insect gaits.

CENTRAL PATTERN GENERATORS

Why do gaits resemble the natural patterns of coupled oscillators in this way? The most likely reason is *not* the mechanical design of animal limbs, although that has to be involved at some stage. Animals use many limb designs, but the range of phase-locked patterns in gaits is much more restricted. The same gait patterns occur in many different animals with very different limbs. Limbs are not passive mechanical oscillators, like a model pendulum, but complex systems of bone and muscle, controlled by equally complicated nerve assemblies driven by signals from the brain. The most likely source of the similarities between nature and mathematics lies in the 'architecture' – the layout – of the circuits in the nervous system that control locomotion.

For a long time, biologists have been convinced that locomotion must involve networks of coupled nerve cells called central pattern generators (CPGs). This belief has always been controversial, because the evidence in its favour is indirect. Nobody has yet dissected out a CPG and shown that it really does control locomotion – except perhaps for the lamprey, an eel-like creature that has no limbs but wiggles through the water using waves of muscular contraction. Nevertheless, the existence of CPGs is very plausible.

It is known that nerve cells can often act as oscillators. So if CPGs really do exist, it is reasonable to expect their behaviour to resemble the dynamics of an oscillator network. There are all sorts of reasons why such a network might be expected to possess symmetry. Many animals are built to a repetitive design, a feature that is especially apparent in centipedes and millipedes but which persists, to some extent, even in quadrupeds. Nearly all animals are left–right symmetric, and it seems likely that CPG design would be left–right symmetric too. And symmetry not only lets us model the phase-locked patterns that are found in gaits; it excludes many other phase-locked patterns that *could* be generated by other models, but are not observed in animals.

It's not only what's there that matters: it's what's not.

Furthermore, the symmetry method solves a significant problem for the CPG theory. Most animals employ several gaits – we noted earlier that horses walk, trot, canter, or gallop. Biologists have often assumed that each gait requires a *separate* CPG. Symmetry-breaking, however, implies that the same CPG circuit can produce all an animal's gaits. Only the strengths of the couplings between neural oscillators need vary.

For all these reasons, it seemed likely that the root cause of phase-locked patterns in animal gaits must be the symmetrical architecture of CPGs. The symmetry that Hildebrand observed in the limbs comes from the symmetry of the networks of nerve cells that control them.

OOPS!

Late in 1996, Golubitsky convinced me that there was a serious technical difficulty with the particular CPG networks that Collins and I had suggested. He'd been worrying about it for years, but had finally pinned down what was bothering him. What it boiled down to was this: if you find one network that supports the walk, the trot, and the pace, then trot and pace will be dynamically equivalent for that network. That is, whenever the network can trot, it can equally well pace. However, horses trot and never pace (apart from some special breeds), whereas camels pace but never trot.

It doesn't fit.

Somehow, we had to separate trot from pace, while not losing the ability to walk. We spent several days drawing various four-oscillator networks on blackboards and equipping the component oscillators with their own symmetries – a trick that we hoped might get round the difficulty.

It didn't.

We even tried making the oscillators 'time-reversible' – meaning that when they were run backwards in time, they behaved just as they did when run forwards. We knew that pendulums were time-reversible, and we had a vague idea that reversibility might help.

It didn't.

Then we came up with a really neat network based on the Möbius band – a strip of paper glued end to end but with a half-twist. Every mathematician knows about the Möbius band, because it has only one side. If you cut a Möbius band along the middle, it doesn't fall into two pieces. When we added a Möbius twist to our network, it looked as if we got exactly the properties that we wanted.

The euphoria lasted one evening, before we discovered a mistake. We had, in effect, counted the symmetries of our Möbius network twice. When we used the right symmetries, the network no longer corresponded nicely to gaits.

At that point, Golubitsky dealt the whole project a severe blow. He proved that *no* network with four oscillators can work. We nearly gave up then – but the *animals* do it, don't they? There had to be a way out.

It was a couple of days later that I tentatively pointed out that we weren't actually obliged to use the same number of oscillators as legs. In fact, a very nice network with eight oscillators looked rather promising. The trick that finally made the idea work – in the sense that it predicted the right phase-locked gait patterns – was to assume that only four of the eight oscillators 'drove' the legs. The other four were there to provide a 'return path' for neural signals, but not to control the limbs. The eight-oscillator network predicted a range of gaits that was very nearly the same as those observed in animals (Figure 53). The main exception was a gait we called the 'jump' – which was like a bound followed by a pause. We stopped worrying about this apparent exception, though, when we were walking through a park in Houston and we saw a squirrel wandering along with precisely that gait.

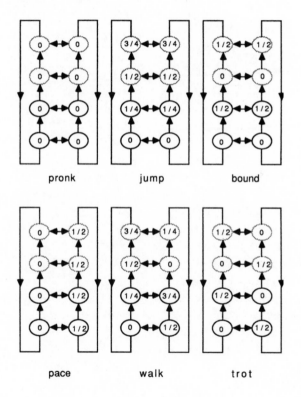

Figure 53   The eight-oscillator network and its primary gaits. Solid ovals indicate oscillators that drive legs, stippled ovals are oscillators that create the return path.

The same kind of circuit works equally nicely for six-legged insects, and even for centipedes. It has a neat 'modular' design that makes it very suitable for mass production in robots. Whether it is what nature actually uses, we have no idea. But it is certainly an interesting guess, and it has most of the right features. We're confident that we're working along the right lines, and our ideas seem to be consistent with those of other workers in the field, such as Nancy Kopell and Bard Ermentrout – who even have good biological evidence for their models☞.

Though it wouldn't totally surprise us to find that nature has some totally different set of tricks up its sleeve.

JUNCTION SIX

*T*he footsteps fade behind you, the ticking of massed arrays of grandfather clocks disappears below the threshold of human hearing. The sounds behind the walls cease. You stop to catch your breath, and look down at your feet. They no longer leave prints on the floor, but in your mind you can see the pattern: left, right, left, right, a double frieze with glide reflection symmetry, laid down in space by the passage of time – and the passage of your feet.

You will always be able to see those patterns, now, whenever you so choose. You will never again watch an animal loping past without looking at the order in which it moves its feet. Everything that moves will leave a luminous trail in your imagination.

But you are not yet out of the maze. Where next?

You have a choice: pass through a low archway to your left, or continue straight on to a corner that turns right.

You continue straight ahead, and turn the corner.

You are standing on a tiny railway platform. The rails run off along the passage, disappearing round the next bend.

At the platform is a train, a single engine. There is no driver, but there is a seat. It looks extremely comfortable. The only controls are a button labelled START and another labelled STOP. You feel that operating such a machine is within your competence, and you are tired from running.

You climb aboard, sit down, and push START.

The train chugs slowly away, down the track, and round the first bend. It comes to a set of points, and takes the left fork. Another track joins from the right, and the points seem set against you! You brace yourself for the crash, but as your train passes through the points they reset themselves automatically to let it pass unharmed. There are more points – hundreds, thousands ... The train trundles on through unmarked tunnels, continually passing through sets of points. Some remain unchanged by its passage, some switch to another setting.

You never meet another train, never pass through a station. You think about pushing the STOP button, but you have no idea how you would ever get out of the maze of tracks. There is no REVERSE button – and even if there were, the points have been reset by your passing.

After a time you wonder whether you're going round in circles. You strain to see some distinguishing feature, a mark on the wall of the tunnel, a chip out of a sleeper. Anything to reassure you that the train isn't caught in an endless loop.

You desperately want to STOP, and think what to do next. You promise yourself that never again will you leap before you've looked.

You wonder whether the train will ever reach a station.

Passage Six

TURING'S TRAIN SET

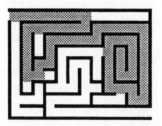

W e are living in an age when computers perpetually seem to get faster, their memories get bigger, their graphics get better, and their price gets smaller. The rate of progress is so great that it is difficult to imagine that there could be any limitations to computers. Weli, there might be limitations such as intelligence or consciousness, but it would be surprising if there were limits to their ability to calculate.

However, such limits do exist.

Ironically, these limits were discovered as a result of an attempt to prove, once and for all, that there are absolutely no limits to mathematics.

In 1900 there was a big international mathematical conference, the International Congress of Mathematicians. The event continues to be held every four years, and nowadays attracts about five thousand participants. In those days the numbers were smaller, but it was still *the* major event in the mathematician's calendar. At the 1900 congress, to mark the new century☞, David Hilbert – a German mathematician generally regarded as the leading mathematician of his time – gave a talk in which he outlined twenty-three major unsolved problems. He also gave a talk on the radio, to a general audience of non-mathematicians.

At that time, by the way – indeed until quite recently – it was entirely normal for the great mathematicians and scientists to popularise their subject. Felix Klein wrote several books on recreational mathematics, and

Hilbert also wrote a popular geometry book with the aid of a co-author. Henri Poincaré wrote popular books on the philosophy and methodology of science. These three were the leading trio of mathematicians at the turn of the nineteenth century. It is only during our present century that science popularisation has somehow become out of bounds for 'serious' academics. Fortunately, that attitude is beginning to disappear again.

Radio was a very new technology in Hilbert's day, and sound recording was in its infancy; nevertheless, a scratchy, noisy recording of Hilbert's talk, made on wire coated in magnetic material, survives. At the end, we can hear Hilbert coming to the climax of his talk, on a ringing note of optimism: 'We must know: we shall know.' Hilbert, having assembled the great unsolved mathematical questions of his time, was affirming his belief that eventually they would all be solved. Indeed, he believed much more: he thought that every question in mathematics must have an answer, and that there ought to be a uniform method for finding it.

He even laid down a programme for discovering such a method – a so-called decision procedure for mathematics. Given such a procedure, Hilbert would then be ready to proceed to the second phase of his programme, which was to prove, beyond any shadow of doubt, that mathematics can never contradict itself. That is, valid mathematical deductions will never lead to a proof that some statement is true, and also to a proof that the same statement is false.

Nobody was really worried that this could happen, but it was a serious philosophical problem. Mathematics is a construct of the human mind: it may *model* reality, but it's not the *same* as reality. Mathematics makes use of idealisations, such as 'infinity', that do not have evident counterparts in the real world. If those idealisations had overreached themselves, then mathematics might contain a hidden, fatal flaw. If such a contradiction were ever to be found, mathematics would collapse in ruins. The reason is that, according to the rules of logic, such a contradiction would imply that *all* statements are true (and also false). For instance, $2 + 2 = 5$ would be true, and so would $2 + 2 = 6$. Fermat's Last Theorem☛ would be true, and Fermat's Last Theorem would be false. The angles of a triangle would not add up to 180º, even in Euclidean geometry. Pythagoras's theorem would be valid for cubes instead of squares. The circumference of a circle would equal its radius ...

The whole exercise would cease to have any meaning.

Probably something could, eventually, be salvaged. A huge effort would be needed to clamber through the rubble, trying to find out what went wrong and how much of the structure could be saved by more careful rebuilding.

If that sounds far-fetched, you should bear in mind that in the mid-1900s there was a major crisis in mathematics, when certain theorems, apparently correctly proved, contradicted other equally plausible theorems. The problems arose in an area known as Fourier analysis☛, in which complicated waveforms are represented as combinations of much simpler trigonometric 'sine' and 'cosine' curves. In those cases it eventually turned out that the mathematicians had been a bit too sloppy, jumping to unwarranted conclusions in various of the more subtle parts of their topic. It taught them to be very cautious indeed – just one of the several reasons why the true mathematician is never happy with 'experimental' evidence, but requires watertight logical proofs.

Extremely watertight.

Hilbert, in fact, was after the biggest piece of watertight logic of them all: a guarantee that there would *never* be another crisis – as long as mathematicians didn't make silly mistakes.

It seemed a reasonable idea at the time. After all, mathematics *feels* perfectly consistent, and all of the evidence is that it has never yet contradicted itself. The equation $2 + 2 = 5$ doesn't exactly ring true. If you've got two pigs in a sty, and you put two more in, you don't – at least not immediately – have *five* fat porkers in the sty.

However, as I've just said, 'experimental' evidence is not the same as proof. So Hilbert devised a line of attack in which mathematics was treated as a purely formal game in which symbols are manipulated according to a system of precisely stated rules. All you had to do then was show that no amount of valid manipulation could ever produce the string of symbols $0 = 1$. It's not so different from proving that in a game of chess there can never be eleven knights of the same colour on the board at once. There *can* be ten: pawns that reach the far side of the board can be promoted to any piece. Queens are usual, but knights are legal. You start with two knights and eight pawns, so if you trade all the pawns for knights … It will never happen in a sensible game, but it's possible. Eleven knights isn't – but can you prove that?

Mathematics is a *much* more subtle game than chess. So subtle that, within just a few years, it was Hilbert's programme that collapsed in ruins. Not because somebody found a contradiction within mathematics, though. Because a young logician called Kurt Gödel showed that Hilbert's programme could never achieve its aims. And, fatally, neither could any similar programme. You could tinker with Hilbert's approach for as long as you liked, but you'd never be able to fix it up.

Even if mathematics *is* consistent, said Gödel, we can never be certain that it is. Indeed, if anyone ever managed to write down a proof that mathematics is consistent, then we would immediately know that it isn't☞.

Gödel's dramatic discovery led to a new understanding of the limits to mathematics, and to formal logic. About thirty years later, in the early days of the computer, before much of the hardware existed and when most computational aids were made from cogwheels or electrical relays, very similar ideas turned out to be fundamental to the theory of computation. In fact, the whole area is much more easily understood if it is viewed from the standpoint of computation.

LOGICAL PARADOXES

At the heart of all mathematics, and of all computation, lies logic. Logic is the glue that holds mathematics together, letting us start from known facts and end up with new ones. Computers are just huge logic boxes, slavishly following prescribed rules to reach desired ends.

Logic is the study of valid deductions. Its raw materials are not numbers, but *statements*. Statements are assertions that are *either* true or false – statements like

dogs bark

or

the United States is a continent.

Here the first statement is true, the second false. However, logic does not concern itself with the factual truth or falsity of specific statements – those fall into other domains, here science and geography. Logic is about the correctness, or not, of deductions of some statements from others. Such as

dogs bark
barking is annoying
therefore dogs are annoying.

According to conventional logic, and leaving aside a few quibbles (such as 'well, *some* dogs don't bark'), this particular deduction is considered to be correct. Indeed, as a *deduction* it is correct even if the individual statements involved might be false – as the second is for many humans. To a logician, the deduction

$$2+2=3,$$
$$3=5,$$
$$therefore\ 2+2=5$$

is entirely unobjectionable; and so is

$$2+2=3,$$
$$3=4,$$
$$therefore\ 2+2=4.$$

In both cases, the logic is impeccable. However, various of the component statements are false, so we are not entitled to deduce that the final conclusion is *true*. As it happens, it is not true in the first case, but it is in the second; however, none of this is related to the validity of the deduction.

Most of the statements that arise in logic are relatively harmless things. One category, however, causes endless problems. This is the category of *self-referential* statements, ones that refer to themselves. The simplest example is

This statement is false.

Let's see: is it true, or is it false? Well, if it's true, then we immediately deduce that it's false. On the other hand, if it's false, then of course it's true. It is a self-contradictory statement, otherwise known as a *paradox*.

The great classical instance of this statement is the alleged claim, by Epimenides the Cretan, that 'all Cretans are liars'. If 'liar' means '*always* tells untruths' then Epimenides is lying when he says he's a liar, so he's truthful, so he *is* a liar after all, so … However, if 'liar' means '*sometimes* tells untruths', as it is usually interpreted, then there is no contradiction. On this occasion he happens to be telling the truth, but maybe on another occasion he won't.

Curiously, no such paradox arises from the statement.

This statement is true.

If it's true, then it's true; if it's false, then it's false. So being self-referential *alone* is not enough to cause trouble.

There are many variants on this 'liar paradox'. One is

There are three mistokes in this semtence.

OK, there are two obvious *spelling* mistakes: 'mistokes' and 'semtence'. But what is the third mistake? Could it be (aha!) that there are actually only *two* mistakes? Yes, but if that's so, then there *are* three mistakes after all, so there are only two, so …

Another variant is a card, on one side of which is the message

> THE STATEMENT
> ON THE OTHER SIDE OF THIS CARD
> IS TRUE

and on the back is another message:

> THE STATEMENT
> ON THE OTHER SIDE OF THIS CARD
> IS FALSE

(Think about it.)

A slightly different kind of logical paradox is the so-called Richard paradox. Some English sentences define numbers – examples are

> The number of planets in the solar system.
> The sum of two and two.
> The number of words in this statement.

Some don't – for instance

> The species to which Eeyore belongs.

Presumably, every statement either defines a number, or it doesn't. So does

> The smallest number that cannot be defined in a sentence of less than sixteen words

define a number, or not? Obviously it does: there are only finitely many words in the English language, there are finitely many sentences with less than sixteen words, so among the numbers they define (when they do) there is a largest one. Add one to that, and you get the number defined by the above sentence.

Fine. So let's count how many words the above sentence has.

Fifteen.

Hmm. We have just defined the smallest number that *cannot* be defined in a sentence of less than sixteen words, using a sentence of less than sixteen words.

Oops!

The classical logicians were forced to grapple with these paradoxes, and find ways to rule them out, otherwise the whole edifice of logic would have come tumbling down.

One way out is simple – and in a sense, is forced upon us. I said that a statement is an assertion that is neither true nor false. For these self-referential assertions, truth or falsity seems self-contradictory. Conclusion: *they are not statements* – not within the meaning of the act. The only problem with this approach is that we can't easily (if at all) tell which sentences *are* statements. Is '2 + 2 = 4' a statement? Well, if Hilbert is right and mathematics doesn't contradict itself, then the answer is 'yes' – but if he's wrong, then the answer is 'no'.

You can't build a sound logical foundation for mathematics if your criterion for what is admissible involves making that kind of choice.

More recently, several unorthodox logical systems have been devised. In one of them, due to Patrick Grim and Gary Mar, the truth of a statement is *dynamic* – it varies over time. So 'this statement is false' is alternately true and false, since each implies the other. In another system, called fuzzy logic, the statement is a half-truth. Literally: it is half true and half false. Despite its name, fuzzy logic is a perfectly precise area of mathematics, and it is the basis of a billion-dollar industry, used in things like washing machines, air conditioners, camcorders, and so forth. Precision is still present because the degree of truth is *precisely* 50%, not 51% or 49%. But rather than developing these esoteric areas of modern logic, we shall make use of the paradoxical nature of certain classical statements to discover limits to mathematical proof and limits to computation.

ALGORITHMS

Nowadays there are thousands of different designs of computer, but they all operate along the same basic lines. They can be thought of as machines that manipulate symbols according to specific rules. The symbols in effect reduce to two, the 'binary' digits 0 and 1. In effect, a computer is a lot of switches with 1 meaning 'on' and 0 'off'. It works by manipulating these switches according to fixed rules.

Symbols more complex than 0 and 1 can be created by stringing lots of 0's and 1's together. The same string of 0's and 1's can have many different interpretations, according to which rules are currently in operation. In arithmetical calculations with whole numbers, a string such as

1001011 is interpreted as the number

$$64 + 8 + 2 + 1 = 75.$$

When manipulating text, it may instead be interpreted as 'ASCII code', in which case it stands for the letter K. The computer doesn't 'know' any of this: it just follows its rules – so it does no serious harm to head towards more familiar ground and think of the symbols manipulated by the computer as being the ordinary numbers 0, 1, 2, 3, 4, and so on. For simplicity, I'll do just that from now on.

The rules take the form of a *program*, a list of instructions that tells you what manipulations to make on the symbols. Real programs are stored in the computer as sequences of symbols 0 and 1 as well, with yet another interpretation – as instructions. Here we can be less formal, and just write down the instructions in words, perhaps like this:

1. Let FRED equal 7.
2. Let ALICE equal 3.
3. Let TOTAL = 0.
4. Add FRED to TOTAL.
5. Subtract 1 from ALICE.
6. If ALICE is bigger than zero, go to step 4.
7. Stop.

Here FRED, ALICE, and TOTAL are names for *variables*, things that can take on numerical values. The idea is to follow the instructions in numerical order, except when an instruction such as number 6 re-routes you.

What does this program do? Let's work through the whole thing, step by step:

1. FRED becomes 7.
2. ALICE becomes 3.
3. TOTAL becomes 0.
4. TOTAL becomes $0 + 7 = 7$.
5. ALICE becomes $3 - 1 = 2$.
6. ALICE is bigger than zero, so go to step 4.
4. TOTAL becomes $7 + 7 = 14$.
5. ALICE becomes $2 - 1 = 1$.
6. ALICE is bigger than zero, so go to step 4.
4. TOTAL becomes $14 + 7 = 21$.
5. ALICE becomes $1 - 1 = 0$.

6. ALICE is not bigger than zero, so continue to step 7.

7. Stop.

We've ended up with TOTAL at 21, ALICE reduced to 0, and FRED still 7. So what this program does is multiply FRED by ALICE and put the result into TOTAL. You can easily convince yourself that with other positive whole numbers for FRED and ALICE the result is again to put the product FRED × ALICE in TOTAL.

This is a typical, albeit very simple and crude, program. Its main feature is a 'loop' from instruction 6 back to step 4, which is carried out when some logical condition (here 'ALICE is bigger than zero') is true, and ignored if that condition is false. It also has the very important instruction *stop*. You can't be sure what a program calculates unless it signals to you that the calculation has *finished*.

Programs don't have to stop. In fact, if you change step 5 to

5a. Add 1 to ALICE,

then the program never escapes from the loop. It keeps going round the steps 4, 5, 6 in turn for ever, with TOTAL increasing by FRED every time, ALICE by 1 every time, and FRED staying exactly the same throughout.

It is relatively easy to spot this particular 'infinite loop' and see that the amended program with instruction 5a never stops. But in the early days of computing it became clear that it's not always so easy to decide whether a program will eventually terminate. Then Alan Turing, a British mathematician with a penchant for logic and one of the two founding fathers of the science of computation (the other being John Von Neumann) discovered that there is no foolproof method of finding out. Not just that nobody had yet found one: that nobody could *ever* find one.

TURING MACHINES

Turing's first step was to find a simple, conceptually 'clean' model of the computational process. Real computers are too complicated; their design involves all sorts of engineering considerations that get in the way of under-standing what they do. Indeed, there are thousands of different makes of computers, and dozens, perhaps hundreds, of different programming languages. However, the differences are not fundamental: they are tactical, so to speak, rather than strategic. In particular, all computers can in principle compute the same things. Indeed, were this not so, some computers would be obviously 'better' than others – able to compute things that others could not – and these would come to dominate the market.

'All computers can in principle compute the same things.' It's an easy thing to say, and it sounds plausible – but is it *true*? How can we prove that two computers can compute the same things? Not by running lots of programs – that's experimental evidence, not proof. No, what we need to do is show that each computer can *simulate* the other (or *emulate*, another term used for the same idea). That is, computer A can run a program which effectively turns it into computer B – in the sense that any program that runs on B can be run in the simulation and yields the same answers – and in the same manner computer B can run a program which effectively turns it into computer A. Simulations normally run more slowly than the thing they are simulating, so for practical purposes A may well be superior to B, say. But if speed is unimportant, then anything A can do, so can B, and conversely.

The way to write a simulation program is to start from the rules of operation of computer B, and program A to answer the question 'in the current circumstances, what would B do, and what would be the result?' Then A can follow, step by step, everything that B would do when running a complex program with large numbers of instructions.

Turing employed this idea, but he took it one stage further. The more complicated your computer's instruction list, the more difficult it is for a mathematical method to analyse what it does. There are just more possibilities to worry about. Mathematicians like to keep things simple – honest they do, even though it may not always appear that way. So Turing developed an extremely simple model of the computational process, known as a Turing machine. It is a 'machine' only in conceptual terms – you *could* build one, with today's technology, but it would be far too cumbersome to use, so nobody bothers to. A Turing machine, slow and cumbersome as it is, is good enough to simulate any existing computer. And the way a Turing machine 'works' is so simple that any computer can easily simulate a Turing machine. In short, a Turing machine and a computer can compute exactly the same things. This is another way to see that all computers can compute the same things: since they all compute exactly the same things as a Turing machine, they obviously all compute the same things as one another☛.

The things they compute, by the way, are called *computable functions*. A *function* is a rule for turning input data (a sequence of 0's and 1's) into output data (another sequence of 0's and 1's); an example is shown in Table 11. The rule is computable if it can be carried out by a Turing

Table 11

input	rule	output
1001110100	Change 0 to 1 and 1 to 0	0110001011
1001110100	Select every second digit	10100
1001110100	Copy the sequence backwards	0010111001
1001110100	Collapse every sequence of successive 0's to a single 0	10111010

machine – or any other computer – by obeying an *algorithm*: a precisely defined sequence of instructions. In short, a program. An algorithm must also be equipped with a guarantee that it will eventually finish the calculation and report its answer.

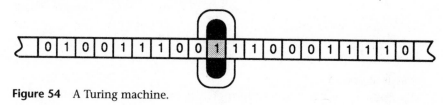

Figure 54   A Turing machine.

For a mental image of a Turing machine, imagine a long tape divided into consecutive square cells (Figure 54). Each cell contains a digit, 0 or 1. You can either use an infinite tape, or if you don't like infinities you just have to be prepared to add some more cells to the tape whenever you need them. The tape travels through a box containing apparatus that can read the digit in a given cell, write another digit in its place if necessary, and move the tape in accordance with various rules. The box itself can be in any of a finite set of internal *states*. For each combination of its state and the digit on the tape immediately beneath it, the box must obey a small list of instructions, like this:

- Leave the current tape digit alone/change it.
- Then move one space left/right.
- Then go into some specified internal state ready for the next step.

Alternatively, the instruction can be just

- Stop

and the computation then finishes.

Here's an example of a typical list of rules, with three internal states, which I'll refer to as states 1, 2, and 3.

- *State 1, Digit 0*: Change digit, move left, go to state 2.

- *State 1, Digit 1*: Stop.
- *State 2, Digit 0*: Leave digit, move right, go to state 3.
- *State 2, Digit 1*: Change digit, move right, go to state 2.
- *State 3, Digit 0*: Change digit, move right, go to state 1.
- *State 3, Digit 1*: Leave digit, move left, go to state 2.

Here, 'move left' and 'move right' refer to the cell of the tape that is in the box. The tape itself moves the other way. In practice, it often helps to think of the box as moving along a fixed tape, in the specified direction.

Suppose we start this machine in state 2, poised over a tape with the digits ... 1001 ... which I'll represent like this:

```
tape   1     0     0     1
cell   ↑
state  2.
```

What does it do?

To find out, we just work through the rules. Tape digits are in bold when they get changed.

- *State 2, Digit 1*: Change digit, move right, go to state 2.
  ```
  tape   0   0   0   1
  cell         ↑
  state        2.
  ```
- *State 2, Digit 0*: Leave digit, move right, go to state 3.
  ```
  tape   0   0   0   1
  cell             ↑
  state            3.
  ```
- *State 3, Digit 0*: Change digit, move right, go to state 1
  ```
  tape   0   1   1   1
  cell             ↑
  state            1.
  ```
- *State 1, Digit 1*: Stop.

So here the program takes the *input* string 1001, turns it into the *output* string 0111, and stops.

This is a short and not terribly interesting computation – but the point is that the machine carries out a definite series of actions on any input string. Sensible programs that compute useful things generally need a lot more states than three. Now, if there are T internal states then, in addition to 'stop' there are $4T$ possible instructions – two choices for the first, times two for the second, times T for the third. You get one Turing machine for each assignment of rules to states.

The digits on the tape provide the computer's *input*; the list of which instructions to perform for which internal state of the box form the *program*, and the list of digits on the tape when the computation stops is the *output*.

The program itself can be written as a sequence of digits 0 and 1, by using some appropriate code – say

000	print 0
001	print 1
010	move right
011	move left
100	stop.

and so on. This is interesting, because it opens up the possibility of *interpreting* a program as data, so that another program can operate on it.

Programs can operate on programs.

THE HALTING PROBLEM

Amazingly, Turing established that these apparently simple devices can carry out any algorithm whatsoever. More than that: there exists a *universal Turing machine*, which can carry out any computation that any other Turing machine can carry out. This one machine can do everything – given the right input.

It has a program, which we'll symbolise as *U*.

For any program *P*, call the corresponding encoded sequence of 0's and 1's code(*P*). The input data for *U* take the form code(*P*) + *d* – that is, the string of 0's and 1's that makes up code(*P*), followed by the string *d*. Here *P* is a program and *d* is the string of input data for *P*. The program *U* scans the first part of its data, the string of the form code(*P*), to find out what *P* would do if it were given data *d*. Then *U* carries out the same action that *P* would on the '*d*' bit of its own input.

In other words, *U* *simulates* the action of *P* on data *d*.

An important feature of any practical algorithm is that it should eventually *stop* – otherwise you don't know whether it's finished working out the 'answer', the output string. The computation in the previous section does stop, but it's also easy to find programs that don't stop. The easiest way to do this is not to have any 'stop' instruction in the list, but there are more subtle ways. In particular, some programs stop for some inputs but not for others.

Is there any way to tell whether a given program will eventually stop with a given input? Suppose you've sat watching the Turing machine churn away, and after ten million steps it hasn't stopped yet. Is there some way you can check whether it's going to stop at all – or must you just keep waiting? Turing named this question the halting problem: for a given machine, can you test which inputs lead to a computation that stops, and which don't? In the spirit of the enterprise, the test ought itself to be an algorithm – so it should be possible to carry it out on some appropriate Turing machine.

Is there a halting algorithm?

In 1936, Turing was thinking about this question, and he wondered what would happen if you applied a putative halting algorithm to a universal Turing machine. More precisely, let U be a universal Turing machine, let its input data be d, and consider the following list of instructions:

- Check whether d is equal to code(D), where D is a program. If not, go back to the start and repeat.
- If $d = $ code(D), double up the data string to give $d + d$.
- If U stops with input data $d + d$, repeat this step indefinitely.
- If U does *not* stop with input data $d + d$, then stop.

If it is possible to solve the halting problem algorithmically, then the third and fourth steps can be carried out algorithmically, so this entire computation can be turned into a program H for a Turing machine.

The construction of H includes exactly two places where the machine may get 'hung up' for ever and not stop. It can do so at the first stage, if d is not the code of a program. It can get hung up at the third stage, if U *does* stop with input $d + d$. These are the only places it can get hung up.

OK, fine. Since H is a program, it has a code, say $h = $ code(H). Now comes Turing's really sneaky question:

Does H halt with input data h?

It certainly doesn't get hung up at the first stage, because the input data h *is* the code of a program – namely, H. The only other place it can get hung up for ever is the third step, and it gets past this provided U does not halt with input $h + h$. That is:

H halts with input data h if and only if U does not halt with input data $h + h$.

So far so good. Now think about how U simulates a program P. It starts with input code(P) $+ d$, and then behaves just like P with input data d. That is:

P halts with input data d if and only if U halts with input data code(P) $+ d$.

This also looks entirely reasonable. But suppose we now let $P=H$, so that code(P) = code(H) = h; and let's also put $d=h$. Then the previous statement becomes

H halts with input data h if and only if U halts with input data $h+h$.

But this exactly contradicts the previous statement but one.

Turing concluded that it is *not* possible to solve the halting problem algorithmically. In brief, his argument is this: if we could solve the halting problem algorithmically, then we could construct a Turing machine that halts if and only if it doesn't.

Turing's proof has a very similar logical structure to the card that I mentioned, whose two sides read

> THE STATEMENT
> ON THE OTHER SIDE OF THIS CARD
> IS TRUE

> THE STATEMENT
> ON THE OTHER SIDE OF THIS CARD
> IS FALSE

In fact, we can summarise Turing's proof as a card that says on one side

> THIS MACHINE HALTS
> IF AND ONLY IF
> THE MACHINE ON THE OTHER SIDE
> HALTS

and on the other side

> THIS MACHINE HALTS
> IF AND ONLY IF
> THE MACHINE ON THE OTHER SIDE
> DOES NOT HALT

The construction of such self-contradictory machines depends on being able to solve the halting problem algorithmically – so Turing deduced that such a solution cannot exist.

THE INTELLIGENT SUBWAY

Turing went on to consider the vexed question of machine intelligence, inventing his famous 'Turing test'. Put a computer in a room and ask it questions from a remote terminal. If you can't tell the difference between its responses and those of an intelligent human being, said Turing, then to all intents and purposes the machine is intelligent.

I'm not going to discuss the merits or otherwise of this proposal, which is often misunderstood. Instead, let me phrase the idea differently. Suppose that the operation of the universe is algorithmic – that is, the 'laws of nature' are a system of mathematical rules, and the universe simply follows those rules from given initial conditions. Then, in effect, the universe is a Turing machine. Its program is the laws of nature, its data are the initial conditions. If an intelligent creature inhabits that universe, then it, too, obeys the laws – so it, too, is (more accurately, can be simulated precisely by) a Turing machine.

In a universe that operates algorithmically, 'strong artificial intelligence' – the construction of truly intelligent machines – *must* be possible in principle. Not necessarily in practice, though, because it may not be possible to encapsulate the rules and the initial data into a sufficiently compact device. For instance, if the 'machine' were the entire universe, we wouldn't accept it as the solution. Indeed, *any* universal Turing machine would be (potentially) intelligent, needing only the right 'intelligence program' and 'intelligence data'.

The real interest in this idea is that is pushes the question of artificial intelligence away from the machine's hardware, and into its program and data.

It also shows that, if we equate intelligence with carrying out some (no doubt complex) algorithm, then a sufficiently complex subway system could become intelligent. It would 'think' rather *s-l-o-w-l-y* ... but it would still be able to 'think' – carry out the intelligence algorithm.

Be that as it may, a subway can certainly compute. Compute what? Anything that a computer can.

It sounds like science fiction. Offhand, I can only think of two science-fictional subway stories. One is A.J. Deutsch's *A Subway Named Möbius*, in which a subway system becomes so complicated that the trains on it cease to have well-defined locations – and so disappear. The other is Colin Kapp's *The Subways of Tazoo*, in his 'unorthodox engineers' series, which chronicles the demise of an alien race that becomes over-reliant on a single kind of power source. But what I have in mind is more like David

Brin's *Earth*, in which the global computer network becomes so inter-connected that it acquires enough intelligence to save humanity from a rogue black hole.

But I'm straying from the point.

In 1994, the journal *Eureka*, published by Cambridge University's student mathematical society the Archimedeans, printed a fascinating article by Adam Chalcraft and Michael Greene◆. It was about the computational abilities of train sets – toy train sets, with rails and points and little men in out-of-date railroad uniforms. They showed that a train set can compute. And whatever a train set can do, a subway surely can.

Agreed, a train set is not quite your Massively Parallel SuperProcessor. Not in speed. But in theoretical computing ability, the two are equally powerful. The theory of Turing machines tells us that any programmable digital computer can simulate any other, given enough memory. Now a computer is just a huge switching circuit with adaptable switches – and trains can switch tracks using points. Suppose, said Chalcraft and Greene, you've got a big enough stock of straight and curved track, bridges, and various kinds of points. However, you've got only one engine and *no* rolling stock. What computations can you perform if you set up the right track layout?

At first it's hard to see how a train can compute at all: it's just a thing on wheels that moves along the rails. But electrons are just things that move along wires, and computers compute using them. The computational aspect is a matter of interpretation in both cases; the underlying mechanisms are much the same. The idea is to encode an input as a lot of 0's and 1's corresponding to the settings of various points in the train layout. Then you run the train along the tracks, and of course those setting change, which in turn alters the path of the train. Eventually the train is side-tracked into a line leading to a terminal – the program 'stops' – and you read off the output from the settings of the same collection of points.

Let's start with the simplest switching unit, known as *lazy points*. They're a Y-shaped piece of track. A train entering the Y from below runs up the upright and out of whichever arm of the Y the points are set for, leaving them set as they were to begin with. However, a train that enters from one of the arms will – if necessary – reset the points so that they connect that arm to the upright, and exit via the upright. But no trains ever come in down one arm and go out via the other (Figure 55(a)). Lazy points have two states, depending on which arm is connected to the upright: let me call them *left* (*L*) and *right* (*R*).

Layouts whose only active components are lazy points have very limited computational ability. Assume that you use a finite number of bits of track,

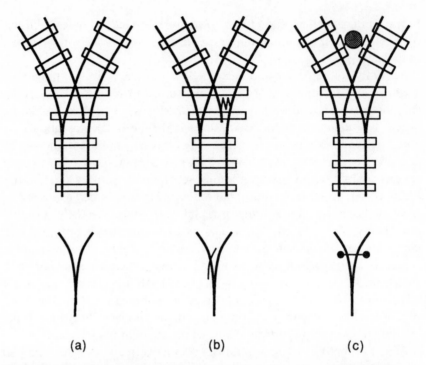

Figure 55   Three types of points (above) and their 'circuit symbols' (below):
(a) lazy points, (b) sprung points, (c) flip-flop.

fix some layout, and let the train run through it. The only things that
change are the points, and these can be in only two states, L or R. If there
are n lazy points, then the layout has at most 2^n states. So eventually the
motion of the train must fall into a repeating cycle, when it starts to run
through a state of the layout that it has run through in the past. Discard
any bits of track that aren't traversed during the cycle. Amazingly, there are
only two topologically distinct types of layout that remain. One is just a
closed loop with no points, and thus computes absolutely nothing at all.
The other is a figure-of-eight arrangements with two points, in which the
arms of each point are joined in a loop and the uprights are run together.

Now, suppose that in the second case the lazy points both start out in
state R, with the train in between them. If you trace its path and the effect
on the points you'll find that it cycles the settings in the sequence
RLRLL**R**L**R**, repeated for ever, where the bold symbols indicate one set of
points and the plain letters indicate the other one. So it's like a computer
that can compute only the terms of the sequence 10100101, repeated
for ever.

For simplicity, I'm letting each *R* represent 1 and each *L* represent 0. I'll clarify how to interpret the train's motion as a computation later, when we get to a more representative layout.

We can now see that we need some more complicated switching gear if we're going to build a computer. The next type of point is a *sprung* one (Figure 55(b)). It's like a lazy point, except that any train entering along the upright of the Y always leaves by the same arm – say the left arm. It doesn't matter which arm you use, because by using bridges you can connect the rest of the layout to either arm, so the right-armed sprung point doesn't do anything new. However, it does simplify layouts to some extent, so I'll allow either arm to be sprung.

The third kind of point is a *flip-flop* (Figure 55(c)) – not standard railroad jargon, but it'll do. With a flip-flop, the train always enters along the upright of the Y, and exits through the left and right arms alternately. Chalcraft and Greene showed that, given these components, we can build a computer – a universal Turing machine.

Here's how.

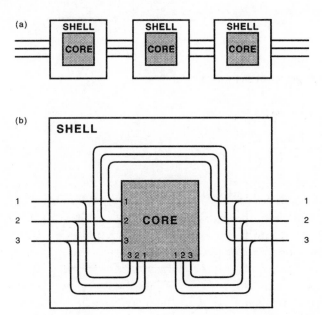

Figure 56 (a) Replacing the tape of a Turing machine by a series of identical track layouts. (b) Each square of the tape consist of a shell, which ensures that trains entering from either direction are treated alike, and a core, which simulates the rules for the Turing machine.

It helps to break the problem down into a series of stages. Recall that a Turing machine employs a box, through which its tape travels. We need to find a train layout that can play the role of the box. Then we just plug an enormous number of these boxes into each other, side by side, to represent the whole tape (Figure 56(a)). Each box will have T tracks coming in from the left and T going out at the right – one for each internal state. Instead of the tape moving through the box, the train moves along the row of boxes. You can tell which 'square' of the 'tape' is being worked on by which box the train is in. But what do we put in the box?

I'll explain how the box is designed in stages. The train tracks are used as both input and output lines, so the box doesn't 'remember' which direction the trains came in from. So it can be set up as an outer shell that feeds trains the same way into an inner core from both input lines, and directs them out again according to the Turing program (Figure 56(b)). Then we can ignore the outer shell, and just concentrate on the design of the core.

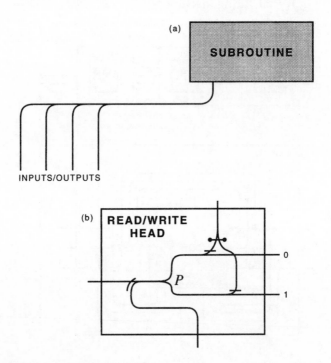

Figure 57 (a) Layout for calling a subroutine: trains enter through lazy points, and exit along the track. (b) Design of a read/write head; note the presence of a flip-flop.

We'll need some gadgets that computer scientists call 'subroutines'. A subroutine is a part of a program that can be used repeatedly by 'calling' it from any other part. You can build complex programs by stringing subroutines together. We can set up a subroutine by hooking up a self-contained sublayout to a whole series of lazy points. Then the train comes in, setting the points as it does so, and wanders round the sublayout until it's carried out whatever subroutine that sublayout computes. Finally it exits by the same track that it came in on, because of the way it set the points on entry. Using one lazy point for each input line, the trains can all be made to enter from the left, carry out the subroutine, and exit to the right along the same track they entered from (Figure 57(a)). We also need one more piece of gadgetry, a *read/write head* (Figure 57(b)). If a train comes in from the left it exits along line 0 or line 1, depending on the

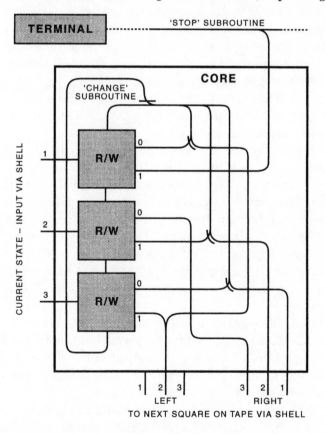

Figure 58 The design of the core circuit: the train enters from the left, obeys the appropriate Turing machine rule, and exits at the bottom.

digit at the 'current' point of the tape; and if a train comes in from above it swaps the 0 and the 1. To achieve this, the lazy point P is set to redirect the train along output lines 0 or 1 according to the digit on the 'tape' at that square. The flip-flop is set so that the first train entering from the top switches P to the other position.

Having got all these bits and pieces, you build the inner core of the box as in Figure 58. You may need some bridges to avoid the tracks crossing, but we can ignore that. The core consists of a parallel set of read/write heads, one for each internal state of the core. The output lines 0 and 1 lead to one of the outputs of the core, or to a lazy point that diverts the train into a subroutine that changes the state of that 'square' on the tape, or to a 'stop' subroutine that guides the train into a single terminal.

In my earlier example, one of the rules is '*State 1, Digit 0*: Change digit, move left, go to state 2'. How does that work?

The square being in state 2 means that the train enters from the side along line 2. This state is set by the output of the *previous* square, which directs the train on to line 2 when it exits. The digit 'written' on that square is 0: that just means that we've set all the lazy points in the read/write heads to 0. So the train comes into the second read/write head, leaves along line 0, and runs into a set of lazy points. These direct it into the 'change' subroutine. It runs vertically downwards, through all the read/write heads, flipping their states from 0 to 1. So the digit 'written' on that square now reads 1, not 0. The train goes back up the vertical track to the left of the heads, exists from the subroutine back on to its original track, and then comes out of the core on the output line *2 left*, which effectively moves the train into the box to the left, in state 2, as required.

Suppose we look at the rule '*State 2, Digit 0*: Leave digit, move right, go to state 3'. The train comes in along line 2 and exits the read/write head along line 0, which leads directly to exit *3 right*. And it never goes anywhere near the subroutine loop, so the state of the square remains unchanged.

And it's equally obvious that the rule '*State 1, Digit 1*: Stop' works properly. Enter along line 1, exit the read/write head along line 1, and you get fed straight into the line that ends at the terminal. By setting up the lines according to the list of rules for the Turing machine, you can make the layout simulate that Turing machine exactly.

These ideas have a fascinating philosophical implication. They show that the future behaviour of a train-set can be undecidable. Turing proved that the halting problem for Turing machines is formally undecidable. Given an arbitrary track layout, you can't predict in advance whether the train is ever going to reach the terminal.

That's quite startling. Most of us have never been terribly bothered about theoretical questions of formal undecidability in computer science. But it's a bit worrying that you could set up a mechanical system with train tracks, whose workings are totally transparent, and not be able to answer such a simple question as whether a train will ever reach a particular station.

LIFE

Nearly twenty years ago, the British mathematician John Horton Conway invented a model of the computational process that is even simpler than Turing's. His original aim was to produce a game, but all along at the back of his mind was the idea that that game should have simple rules but very complex results – as complex, indeed, as any computation.

Conway called his game Life☞, and it will give you some very useful intuition about how simple rules can lead to complex behaviour. Human beings have a habit of assuming that the complexity of behaviour must somehow be present in the causes of that behaviour – that is, in the rules that generate it. Life is one of a number of mathematical gadgets that demonstrate that this intuition is false. Complexity can be generated *spontaneously* by very simple rules. Not by all simple rules, though: the rule 'do nothing' is simple, but what it generates is totally static and boring. Some simple rules lead to complex behaviour; others do not. One of the morals of this chapter is that we should not expect to find an *easy* test for which rules do or do not generate complexity.

Life is a cellular automaton, a concept we've already met. A huge chessboard, with cells that change colour according to some fixed system of rules. Life uses only two colours, black and white, and there are only three rules:

- A cell that is white at one instant becomes black at the next if it has precisely three black neighbours.
- A cell that is black at one instant becomes white at the next if it has four or more black neighbours.
- A cell that is black at one instant becomes white at the next if it has one or no black neighbours.

In all other cases, cells maintain their colour. The neighbours of a given square are the eight cells adjacent to it vertically, horizontally, or diagonally. All changes are deemed to be made simultaneously.

The idea is to start with an object made up of black cells, and the rest of the board white. Then you follow the rules and watch how that object changes. For example, a 2×1 block dies out at the first move. A 2×2 block doesn't do anything, so it survives indefinitely. More interesting is a simple shape called a glider: it moves. It changes shape in a cycle of length four, and by the end of the cycle it has moved one cell diagonally. More complicated shapes, spaceships, move horizontally or vertically. A 'glider gun' which changes through a fixed cycle of 30 shapes fires an endless stream of gliders (Figure 59).

Life is a rule-based universe. The future of any initial state is completely determined by the three rules – Life's 'Theory of Everything'. But in practice it may be very hard to predict what will happen, even though it's all implicit in the rules. To make it explicit, you have to follow the rules and see what happens: there don't seem to be any short cuts. Big shapes can collapse, small ones grow, and there are always surprises. Conway proved that the outcome of the game is inherently unpredictable, in the sense that there is no way to decide in advance whether a given object will survive indefinitely, or disappear entirely. He did this by constructing a 'programmable computer' that uses pulses of gliders instead of electrical

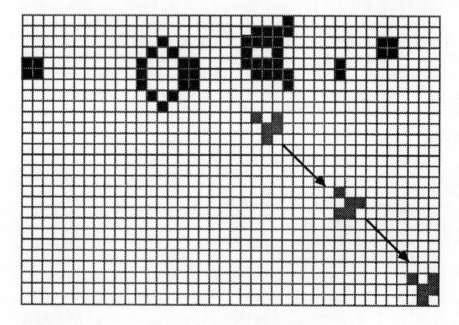

Figure 59   Glider gun (black) fires a stream of gliders (grey).

impulses to carry and manipulate information. Turing's work shows that there cannot exist a computer program that can decide in advance whether any given program, when run on a given machine, will go on for ever, or will stop. The only way to find out is to run the program and watch. If it stops, you know; if it keeps going, you have no idea whether it will continue to keep going, or whether it's just about to stop as soon as you give up and go away.

You can run such an undecidable 'program' on the Life computer. You can even arrange matters so that *if* the program stops then the entire structure annihilates itself by shooting itself down with gliders. So either it goes on for ever, or it wipes itself out. The problem is that there's no way you can tell which. The long-term fate of this particular object, in Conway's simple rule-based game, cannot be determined in advance by any finite computation. We can only let the system run, and watch what it does.

The universe may also be like this, deep down. Even if we knew the rules by which the universe operates – *its* Theory of Everything – we still might not be able to decide what the implications of those rules were.

GÖDEL'S THEOREM

A problem like the halting problem, for which there exists no algorithmic solution, is said to be *undecidable*. Turing's problem was not the first to be proved undecidable, but it is in many ways the simplest example of such a problem. The first problem to be proved undecidable was a mathematical one. It had a rather similar 'feel' to it, but instead of programs and input data it worked with mathematical theorems and proofs. This was the aforementioned work of Kurt Gödel, which demolished Hilbert's programme to put all of mathematics on irrefutably sound foundations.

Gödel must at some point have convinced himself that the Hilbert programme was too ambitious ever to succeed. Indeed, when you come to think about it, a mathematical proof of the infallibility of mathematics looks like circular logic. For example, if mathematics is logically flawed, then even if you could write down such a 'proof', it could well be fallacious – because the language in which it is written is logically flawed. Hilbert was aware of this difficulty, but he convinced himself that it was possible to get round it by viewing mathematics in two distinct ways: as a meaningful series of logically related statements, or as a manipulative game played with meaningless symbols. Gödel found a way to pin down the feeling that the logic of Hilbert's programme was circular. At its heart – just as with Turing's work – there was a logical paradox.

Turing's proof of undecidability for the halting problem is modelled on the card whose two sides bear contradictory messages. Gödel's theorem about the undecidability of arithmetic is similarly modelled on the even simpler 'liar paradox':

> This sentence is false.

Instead of truth and falsity, however, Gödel worked with provability. A mathematical statement is said to be *provable* if there exists a proof – starting from some formal system of logical axioms for mathematics, and deduced using valid logic. If no such proof exists, then the statement is said to be *unprovable*. Provability is a concept well suited to the metaphor of the magical maze. The axioms for mathematics are the entrance to the maze. The rules of logic take us along passages to new junctions, new logical statements – new theorems. A theorem is provable if we can find a path through the maze that leads to it, starting from the entrance. It is unprovable if there is no such path. The unprovable theorems lie in inaccessible regions of the magical maze.

The natural assumption for Hilbert and his predecessors was that unprovable is the same as disprovable. A theorem is disprovable if its negation is provable: for instance, by proving that $2+2=4$ we disprove $2+2 \neq 4$. A theorem is disprovable if we can find a path through the maze that leads to the theorem's *negation*, starting from the entrance. If mathematics is consistent, as Hilbert was sure it must be, then it is impossible to find a path through the maze that leads to a theorem, and another path through the maze that leads to its negation. The provable theorems are true, the disprovable ones are false, and nothing can be both. This implies that all disprovable theorems lie in inaccessible regions of the magical maze: anything disprovable is certainly unprovable.

Hilbert thought that the two concepts ought to be identical.

Suppose, for the sake of argument, that Hilbert is right. Then every mathematical statement is either provable or unprovable – and not both. (In the jargon, mathematics is both *complete* and *consistent*.) Then a statement will be provable if and only if it is true, and unprovable if and only if it is false. We here assume the axioms to be true, hence any valid logical consequence of them is also assumed to be true. If a statement S is true, then not-S is false; if S if false, then not-S is true. The magical maze of all possible mathematical statements divides into two 'mirror image' pieces – a white piece of true and provable statements, and a black piece of false, unprovable, and inaccessible statements☞. The mirror that maps one piece to the other is the logical operation 'not'. The pieces do not overlap (consistency), and there are no 'grey areas' that are neither black nor white (completeness).

It's a comfortable picture, but Gödel showed that it is an incorrect one. He proved that either black and white areas must overlap (and so mathematics is inconsistent) or there have to be some grey areas (and some statements can neither be proved nor disproved). His idea was to make use of the simplest logical paradox, by constructing a mathematical theorem that states:

> This theorem cannot be proved.

He couldn't do this directly – 'this theorem' has no obvious interpretation as a statement in formal mathematics. So he played Hilbert's game of interpreting mathematics in two different ways; and he played it to Hilbert's detriment. The first step, like Hilbert's, was to think of any mathematical statement as a string of symbols, so that '$2+2=4$' is no longer considered as a statement about arithmetic – one that must be either true or false, and is actually true. Instead it is just a string of five meaningless symbols:

$$\text{'2'} \quad \text{'+'} \quad \text{'2'} \quad \text{'='} \quad \text{'4'}.$$

Not only statements, but entire proofs, are strings of symbols.

Next, Gödel found a way to encode any symbol string as a number – and conversely to decode numbers into symbol strings. This is a bit more tricky, but here's one way to do it. First, assign a fixed numerical value to each symbol – perhaps as in Table 12 (the sequence can be continued indefinitely). It doesn't matter how you do this, but the Gödel code has

Table 12

symbol	value
–	1
=	2
1	3
2	4
3	5
4	6

to be set up at the start and never changed thereafter. Then you take a string of symbols, such as 2, +, 2, =, 4, and replace it by the corresponding sequence of code values 4, 1, 4, 2, 6. This is a whole series of numbers, not just one, but now you calculate the single number

$$2^4 3^1 5^4 7^2 11^6,$$

where 2, 3, 5, 7, 11, … are the primes, and the powers 4, 1, 4, 2, 6 are the code values for the symbols in the original string, in order.

The result is a pretty big number, even for a short string of symbols, but that doesn't matter. What is important is that each symbol string determines a single number, and conversely each number determines a symbol string. Going from symbol strings to numbers is straightforward and algorithmic. Going the other way – decoding the number to get its symbol string – relies on the fact that a number can be represented as a product of primes in exactly one way. So if we start with a number such as 4800, we find its symbol string like this:

- Start with 4800.
- Write it as a product of primes: $4800 = 2^6 3^1 5^2$.
- Look at the sequence of powers: 6, 1, 2.
- Write down the corresponding symbols: 4, +, =.
- Remove commas to get the symbol string: $4 + =$.

It may not be a meaningful string – here it isn't – but we know which string it is.

What is a proof? It is a series of logical manipulations on symbol strings. By way of the Gödel code, it can be converted into a series of manipulations on *numbers*. Those manipulations can be represented in terms of ordinary arithmetic – not easily or obviously, but in a straightforward way once you adopt the right viewpoint. So the existence of a proof of some statement can be rephrased in arithmetical terms. Instead of

Starting from the axioms, there exists a series of logical deductions that leads to statement X,

we have the equivalent statement

Starting with this list of numbers and applying one or more of the following arithmetical processes, there exists a series of logical deductions that leads to the number that is the code for X.

This second statement provides an arithmetical interpretation for 'there exists a proof of'. In particular, Gödel could give an arithmetical interpretation of statements such as

> There exists a proof of the symbol string with Gödel code 1066,

or

> There does not exist a proof of the symbol string with Gödel code 1066.

These statements, too, could be represented by single numbers.

Finally, he arranged matters so that the Gödel code for 'There does not exist a proof of symbol string 1066' was the *same* number, 1066, that was referred to in the statement itself. Of course he didn't use 1066 as such, but for simplicity we can pretend that that was the number. The actual number was *huge* – too big to write down, but not too big to specify indirectly.

In effect, symbol string 1066 said (in a suitable interpretation) that 'this statement has no proof'. It was – via its code – self-referential.

Now, suppose that we are in the situation that Hilbert considered desirable – and achievable – in which 'provable' is the same as 'true', and 'unprovable' is the same as 'false'. Then is statement 1066 true or false?

If statement 1066 is true, then since it asserts that statement 1066 is unprovable, it follows that statement 1066 is false.

If statement 1066 is false, then it is unprovable, so the statement 'statement 1066 is unprovable' is true. But this is statement 1066, so statement 1066 is true.

If statement 1066 is true, then it is false; similarly if statement 1066 is false, then it is true.

Notice that Gödel is *not* asserting that such a statement actually exists. What he is asserting is more subtle: *if* Hilbert is right that 'true' and 'provable' are the same thing, *then* such a self-contradictory statement exists. Since self-contradictory statements cannot occur in a logically consistent mathematics, we then deduce that *either*

> Hilbert was wrong

or

> Mathematics is logically inconsistent.

The second statement is just another way of saying that Hilbert was wrong, but for a different reason. So either way, Hilbert was wrong.

He wasn't very pleased to be told this. But he was a good enough mathematician, and an honest enough person, to recognise that Gödel was right.

'We must know, we shall know,' said Hilbert in his radio broadcast. Gödel knew better. There are some things that we *cannot* know.

JUNCTION SEVEN

*T*he train pulls into a station, and halts. You do not even have to push the
STOP button. You sag into the driver's seat and draw a deep breath.

Ahead are buffers. The station is a dead-end. You took the wrong path,
wandered over to the wrong side of the railway tracks, and discovered only a
blank wall, blocking any further progress through the magical maze.

The STOP button has disappeared. In its place is one marked REVERSE. You
hesitate, but only for a moment. What other choice is there? You push the
button. The train chugs gently backwards, out of the station. After a worryingly
long sequence of twists and turns, and much resetting of points, as always, you
return to where you went astray.

Nervously you jump out of the engine, but it just sits there – waiting for the
next foolish passenger.

You know where to go, now. You turn sharply to your right, through an archway, into the seventh passage of the magical maze. Over the arch is engraved a
motto:

THINK FIRST, CALCULATE LATER

The archway leads to an irregular, zigzag passage, with several blind alleys
branching off … you almost become trapped again, but you realise your error
and backtrack – fast. A dozen burly workmen are digging trenches. Some are
feeding lengths of cable into the trenches from large reels. Others are pulling it
out again – and tempers are becoming frayed. A foreman, consulting a large
plan, scratches his head in bewilderment. You tiptoe quietly past them.

A tiny bubble drifts past on the breeze. Then another, and another. Soon they
are coming thick and fast, shimmering in all the colours of the rainbow. There
are bubbles as big as your head, as big as your body – one bubble so huge that
you could imagine living in it.

There are bubbles joined to bubbles, bubbles inside bubbles, long strings of
bubbles like plump, round sausages …

You follow the trail of bubbles towards its source.

Now the passage is beginning to fill with foam. It rises to your ankles, your
knees, your waist …

You decide it's time to get out of here.
By whichever route is the quickest.

Passage Seven

QUEEN DIDO'S HIDE

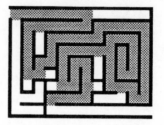

Why are soap bubbles spherical?

Some aren't, actually. A moving bubble is often irregular in shape. But a bubble that floats very gently on the breeze is generally a sphere. The reason is energy.

The energy in a film of soap depends on its area. The smaller the area, the smaller the energy. Nature is fundamentally lazy, and does everything using the least energy possible. So a soap film always has the smallest area possible, consistent with doing its job. The job of a bubble is to contain a given quantity of air. The surface of smallest area that contains a given quantity of air is a sphere.

That's why soap bubbles are round.

One of the things that mathematics is good for is to help us find which thing, of a given kind, is the longest, the shortest, the best, the biggest, the smallest, or the cheapest. What shape should a piece of card be in order to make the largest box? You can imagine that a company which sells groceries, say, would be interested in such questions. They might find the answer 'empirically', by trial and error; but if they were producing goods in large enough quantities they would prefer to have some kind of mathematical guarantee that they weren't wasting cardboard by employing a poor design. Similarly, an airline company wishing to run services between a number of cities would be interested in finding out how to schedule the flights so as to maximise their profits.

Problems of this type come under the general heading of 'optimisation' – finding the best solution. Like all mathematics – and science, and technology, and medicine, and business, and just about everything else – the objective is not to find the absolute best solution to the problem among all possible ways of tackling it, but to find the best solution subject to specific assumptions about what kind of answer we're looking for. That is, we always work in a chosen context. The *absolute* best solution to the airline's wish to maximise its profits, for example, might be to sell up and buy shares in a bank – but that goes outside the context of finding the best schedule of flights.

I mention this point for three reasons. First, it sheds light on how the mathematical mind goes about its work: unless a specific context can be formulated, mathematics really can't be brought to bear on a problem. Second, I want you to be aware of the context of our investigations. Third, I don't want you to think that mathematics can somehow be used to *prove* that something in the real world is the best possible thing there is. You might think that nobody would try to argue in that manner, but in fact many people do it all the time. For example, one of the arguments that politicians often advance in support of 'free market' economics is that we *know*, mathematically, that this always leads to the best possible outcome. Now, it is indeed true that there exists a theoretical proof of the optimality of the free market: it is taught in every undergraduate economics course. However, this 'proof' rests on a whole series of assumptions – that all agents have perfect information abut the state of the market, that supply and demand are related by perfect mathematical rules, and so on. These assumptions are standard in classical mathematical economics. Nevertheless, the real economy does *not* satisfy those assumptions, so the mathematical model cannot legitimately be used to argue that the free market is best in any absolute sense.

In fact, it's not terribly clear in this case that it is best in any sense, except for serving the short-term interests of a small number of wealthy businessmen. The contextual questions that need to be asked include 'best for whom?' and 'best at achieving *what*?' The mathematics cannot answer those questions: instead, it begins from the assumptions that arise when you choose a particular answer to them.

I also don't want you to be misled about how the mathematical techniques of optimisation would be used in real life. Carrying out the necessary calculations on serious practical problems is a complicated task, involving a mixture of general mathematical principles, computer calculations, and some 'seat of the pants' adjustments to take account of features that the mathematics can't easily handle. For instance, you might come up with a wonderful airline schedule that has one tiny flaw: one flight has to land at an airport somewhere in the early hours of the morning, when flights are banned because of local laws to prevent noise. So don't expect to emerge from the magical maze with an amazing new idea about the design of cardboard boxes that will make you a fortune with the supermarket companies.

DIDO'S HIDE

The earliest recorded example of a mathematical solution to an optimisation problem is the ancient Greek legend of Queen Dido. Dido, it is said, was given a bull hide and told that she could take possession of whatever land she could enclose with it. Being rather bright, she cut the hide into an enormously long, thin strip, and arranged it in a huge circle, thereby enclosing the largest possible area of land. On it, she founded the city of Carthage.

When we start to analyse this question more critically, you'll see what I mean by choosing a context. What if Dido had been able to cut the hide into a thinner strip still? Then Carthage could have been a lot bigger ... But it's hard to see how to formalise the question 'how fine can you cut bull hide?' – and if you can't do that, then there's nothing to get your mathematical teeth into.

Such is the art of mathematical modelling.

Dido found the answer to the two-dimensional version of the soap bubble problem – to enclose a given area with the shortest curve. Our job is to prove that her answer was right. To do the same for a three-dimensional bubble is beyond our powers – it *can* be done, but only with a lot of mathematical technique. Even the two-dimensional version is far from easy.

Instead of frittering away hours on fruitless speculation about the thinness of strings made from a bull's hide, we'll assume that we are presented with a *fixed* length of hide, and ask whether a circle is indeed the largest area that it can enclose. We'll also model the long, thin hide by a mathematical curve – which in principle has zero thickness. You can't do that exactly with bull hide. No matter: if the hide is enormously longer than it is thick, we won't go too far wrong by simplifying the analysis in that way. And now we've got a problem that can be subjected to sensible mathematical analysis. Given a curve of fixed length, what shape should it be to enclose the greatest area?

It took mathematicians quite a long time to realise that questions like this come in two parts:

- Show that an answer to the question *exists*.
- Find out what it is.

Answers to mathematical questions do not always exist, even if the question looks reasonable. Here's an example. It is known that the shortest path between two given points (in the plane) is a straight line. We might ask 'what is the shortest *non-straight* path between two given points?' In effect, this asks what the next shortest path, after the straight line, is. However, there is no such beast. Given any path that's not straight, you can always find a shorter path that is also not straight, by taking a short cut across some bend. So the shortest non-straight path *does not exist*.

In Queen Dido's problem, it turns out that a solution does exist. In 1838 the great geometer Jacob Steiner found a beautiful argument to show that *once you know that an answer exists*, you can see that it has to be a circle. His basic idea is that if you take any curve that is not a circle, then you can change it to increase that area that it contains. This is a line of attack that will be familiar to Sherlock Holmes fans: 'Once you have eliminated the impossible, then whatever remains, however improbable, must be the truth.'

However, the great Holmes would immediately have understood that before you find a murderer by eliminating everybody else from your enquiries, you had better be very sure that a murder has been committed. Otherwise you, dear reader, could readily be convicted of the murder of Toad of Toad Hall. Did your next door neighbour do it? No, they have an alibi. Did the grocer do it? No, she has never been near Toad Hall in her life. Did President Bill Clinton do it? Of course not, he was in the White House at the time. And so on. If enough enquiries are made – five billion should do it – the *only* suspect left is you.

By Holmes's principle, the proof of your guilt would be watertight – were it not for the awkward fact that that Toad is a fictional character, so there never was a murder. Steiner was in the position of pinning down the murderer's identity *provided a murder had been committed*, but he never actually produced a corpse. He never seemed able to understand that his proof was incomplete, though in 1842 he half-acknowledged the difficulty in print.

Without a proof of existence, Steiner's style of argument can lead to fallacies. Not just in the context of murders, but in mathematics too. Probably the simplest such fallacy is this statement:

The largest non-zero whole number is 1.

If Steiner's line of argument – ignoring the question of existence – is valid, then we can easily establish that 1 is the largest non-zero whole number. To do so we merely show that, given any non-zero whole number that is not equal to 1, we can find a bigger one. That is, no non-zero whole number different from 1 can be the largest. So the only possibility left is 1 itself. To do this is easy. If you take any non-zero whole number that is not equal to 1 and multiply it by itself, you get something even bigger. *Only* for the number 1 does this step not produce a bigger number, because $1 \times 1 = 1$. Conclusion: any non-zero whole number other than 1 cannot be the biggest non-zero whole number. But does this permit us to conclude that 1 is the biggest? No. It's the Toad syndrome in action. All we can legitimately conclude from the argument is that *either* 1 is the biggest, *or* the biggest does not exist. Of course in this case we know, on other grounds, that it is the second statement that is true: there is no biggest non-zero whole number. Proof: if there were such a number, then it would be greater than or equal to any non-zero whole number – for example, itself plus one. But *no* number is greater than or equal to itself plus one.

Steiner would have seen the fallacy in this numerical argument, but he didn't seem able to grasp that his own 'proof' for Queen Dido's problem suffered from the same potential fault. The difference, perhaps, was that on this occasion his answer was correct. It is so intuitive that the answer is a circle that it's hard to consider the possibility that no answer exists. But in mathematics, intuition and proof are not the same – and you may get the right answer by faulty reasoning, which is what Steiner did.

CIRCULAR REASONING

Other mathematicians quickly filled the gap in Steiner's proof, however, by showing that in this case an answer does exist. Bearing that in mind, we can now appreciate how clever the rest of Steiner's proof was. It goes like this.

Suppose we have, by some method, laid hands on a curve of the chosen length that really does contain the largest possible area. We will now infer, from the maximal area property, various other features of that curve. The aim is to show that it has to be a circle, and we'll reach that deduction in several easy stages. Here's the first step:

Step 1: The curve is convex.

By 'convex' I mean that, given any two points inside the curve, the line segment that joins them also lies inside the curve. In other words, there are no 'dents' where the curve bulges inwards as in Figure 60(a). Well, suppose there is such a dent. Then we can find a line that touches the curve at two points, as in Figure 60(b). Then we form a new curve by reflecting part of the old one in a mirror that lies along that line, as in Figure 60(c). Clearly the resulting curve has the same length as the old one – because the length of the reflected part doesn't change. However, the new curve encloses a larger area – the shaded part in the figure. But the original curve enclosed the *largest* possible area! The only way out of the logical impasse is that no such enlargement is possible. The inevitable conclusion is that the original curve must have been convex all along.

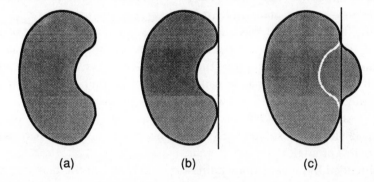

(a)   (b)   (c)

Figure 60   If there is a dent (a), find a line that touches the curve twice (b) and reflect (c) to increase the area without changing the length.

An important feature of a convex curve is that if a line cuts across it, then it divides the region inside the curve into exactly *two* parts. Steiner needed this property for his second step. Before describing how that goes, it's useful to have some terminology. Say that a line is a *diameter* of the curve if it divides the perimeter into two equal parts.

Step 2: Every diameter divides the *area* into two equal parts as well.

Suppose that some diameter does *not* divide the area into two equal parts. Then we can take the piece with the larger area (Figure 61(a)), reflect it across the line (Figure 61(b)), and thereby create a new curve (Figure 61(c)) with the same perimeter as before, but larger area. Again,

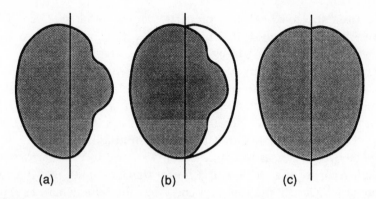

Figure 61   If some diameter does not divide the area equally, take the bigger piece (a), reflect it (b), and get a curve (c) with the same length but bigger area.

since the original curve enclosed the largest possible area, no such enlargement is possible. Therefore the assumption that some diameter does not divide the area into two equal parts must be false. Therefore every diameter divides the area into two equal parts, as required.

With this established, we can simplify the problem by looking at just half of the curve. Choose some diameter, and cut the curve in half. Its area also halves. If we can prove that this half-curve must be a semicircle, then it follows – by doing the same for the other half – that the whole curve must be a circle. And that's how Steiner proceeded. First he proved a very neat geometrical property:

Step 3: The angle subtended (in the halved curve) by its diameter is always a right angle.

'Subtended by' means draw a line from each end of the diameter to a point on the curve: then see what angle they meet at. Suppose that the angle subtended by some diameter is not a right angle. Then it is either more than a right angle, or less than a right angle. If it is less than a right angle, we can increase the area of the curve☞ by 'spreading the angle out' as in Figure 62(a). The same goes if the angle is more than a right angle (Figure 62(b)), except that now you have to narrow the angle instead of opening it out.

Step 4: The curve must be a circle.

Choose a diameter, and let its ends be A and B. Choose any point C on the curve. We know that angle ACB is 90º. It is a general theorem in geometry that the angle in a semicircle is 90º. Less well known, but also true, is the converse: if all such angles are 90º, then the curve is a semicircle☞.

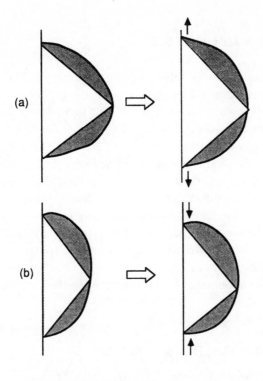

Figure 62   (a) The angle subtended by a point on a curve. (b) If the angle is less than 90º we can spread it out and increase the area. (c) If the angle is more than 90º we can squash it together and increase the area.

Step 5: Go for the jugular.

Since each half-curve is a semicircle, and they adjoin along a common diameter, the whole curve is a circle. Done!

Neat.

Admittedly the argument is a bit long-winded, though by the standards of professional mathematics it's actually pretty *short*. Moreover, a lot of it could be made shorter still for a professional, because they will have met similar ideas elsewhere.

As I've said, Steiner's geometrical fun and games does not actually prove that the circle is the curve of maximal area for a given perimeter. It proves something subtly different – and weaker: *if* there is a curve of maximal area for a given perimeter, then that curve has to be a circle.

If anybody killed poor Toad, it must be *you*.

ARITHMETIC AND OLD LACE

Our next optimisation problem is about shoelaces. The potential for extracting significant mathematics from shoelaces was not widely recognised until it was noticed by John Halton☛. There are at least three common ways to lace shoes, shown in Figure 63: American zigzag lacing, the European straight lacing, and quick-action shoe-store lacing. From the point of view of the purchaser, styles of lacing can differ in their aesthetic appeal and in the time required to tie them. From the point of view of the shoe manufacturer, a more pertinent question is which type of lacing requires the shortest – and therefore cheapest – laces. Here I'll side with the shoe manufacturer, but you might care to assign a plausible measure of complexity to the lacing patterns illustrated, and decide which is the simplest to *tie*.

Of course, the shoemaker is not restricted to the three lacing patterns

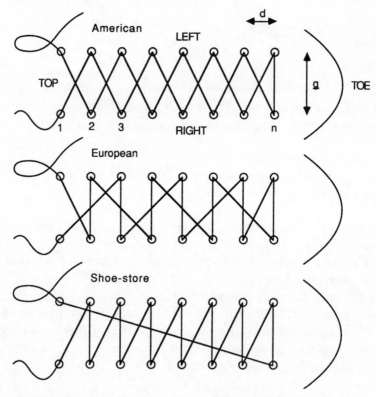

Figure 63   Three common shoelace patterns. Which uses the least lace?

shown, and we can ask a more difficult question: which pattern of lacing, among *all* the possibilities, requires the shortest lace? Halton's ingenious methods answer this too.

To keep the discussion simple, I'm going to assume that the lace moves alternately from the left row of eyelets to the right and back again. Some perfectly practical ways to lace shoes don't do that, and some of them are shorter than anything I'm going to describe here. I'm choosing my context and I'm sticking to it – and my conclusions will be valid only with that context. I'll focus only on the length of shoelace that lies between the 'top' two eyelets of the shoe, on the left of the diagrams – the part represented by straight line segments. The amount of extra lace required is essentially that needed to tie an effective bow, and is the same for all methods of lacing, so it can be ignored.

My terminology will refer to the lacing as seen by the wearer (hence 'top' just now), so that the upper row of eyelets in the figure lies on the left side of the shoe, and the lower row on the right. I shall also idealise the problem so that the lace is a mathematical line of zero thickness and the eyelets are points. Using a brute force attack, the length of the lace can then be calculated in terms of three parameters of the problem:

- The number n of pairs of eyelets.
- The distance d between successive eyelets.
- The gap g between corresponding left and right eyelets.

With the aid of Pythagoras's theorem (one wonders what the great man would have made of this particular application), it is not too hard to calculate the lengths for the lacings in Figure 63☞. Suppose, for the sake of argument, that $n=8$, as in the figure, $d=1$, and $g=2$. Then the lengths are:

American:	37.777.
European:	40.271.
Shoe-store:	42.134.

The shortest is American lacing, followed by European, and finally by shoe-store. But can we be certain that this is always the case, or does it depend on the numbers n, d, and g?

Some careful algebra shows that if d and g are non-zero and n is at least 3, then the shortest lacing is always American, followed by European, followed by shoe-store. If $n=2$ and d and g are non-zero, then American is still shortest but European and shoe-store lacings are of equal length. (If $n=1$, or $d=0$, or $g=0$, then all three lacings are equally long, but only a mathematician would worry about such cases!) However, the algebraic approach is complicated, and offers little insight into what makes different lacings more or less efficient.

Instead of using algebra, Halton described a clever geometrical trick which makes it completely obvious that American lacing is the shortest of the three. With a little more work and a variation on that trick, it also becomes clear that shoe-store lacing is the longest.

FERMAT'S PRINCIPLE

Halton's idea owes its inspiration to optics, the study of the paths traced by rays of light. Mathematicians discovered long ago that many features of the geometry of light rays can be made more transparent – if that is the word to use when discussing light – by applying carefully chosen reflections to straighten out a bent light-path, making comparisons simpler. For example, to derive the classical law of reflection – 'angle of incidence equals angle of reflection' – at a mirror, consider a light ray whose path is composed of two straight segments: one that hits the mirror, and one that bounces off. If you reflect the second half of the path in the mirror (Figure 64), then the result is a path that passes through the front of the mirror and enters Alice's mirror-world behind the looking-glass. According to the principle of least time, a general property of light rays enunciated some centuries back by Pierre de Fermat, such a path must reach its destination in the shortest time – which in this case implies that it is a straight line. Thus the 'mirror angle' marked in the figure is equal to the angle of incidence – but it is also obviously equal to the angle of reflection.

Figure 65 shows geometric representations of all three types of lacing,

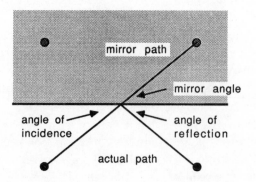

Figure 64 Fermat's principle determines how a light ray reflects off a mirror.

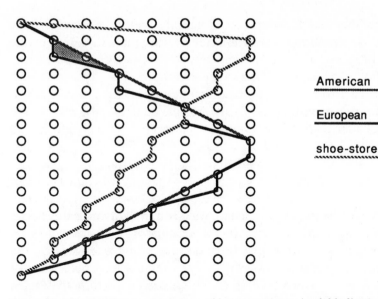

American

European

shoe-store

Figure 65   Geometric representation of lacing patterns, 'unfolded' using mirrors.

which Halton derives by an extension of this optical reflection trick. The figure requires a little explanation. It consists of $2n$ rows of eyelets, spaced distance d apart in the horizontal direction. Successive rows are spaced distance g apart vertically, and in order to reduce the size of the figure we have now reduced g from 2 (as it was in Figure 63) to 0.5. The method works for any values of d and g, so this causes no difficulty. The first row of the diagram represents the left-hand row of eyelets. The second row of the diagram represents the right-hand row of eyelets. After that, rows alternately represent the left-hand eyelets and the right-hand eyelets, so that the odd-numbered rows represent left-hand eyelets and even-numbered rows represent right-hand eyelets.

The polygonal paths that zigzag across this diagram correspond to the lacings, but with an extra 'twist' – almost literally. Start at the top left eyelet of a lacing pattern and draw the first segment of lace, running from left to right of the shoe, between the first two rows of the diagram. Draw the next segment of lace *reflected* to lie between rows 2 and 3, instead of going back from row 2 to row 1 as it does in a real shoe. Continue in this manner, reflecting the physical position of each successive segment whenever it encounters an eyelet. After two such reflection, the segment will be parallel to its original position but two rows lower, and so on. In effect, the two rows of eyelets are replaced by mirrors. So, instead of zigzagging between the two rows of eyelets, the path now moves steadily down the figure, one row at a time, while its horizontal motion along the rows repeats precisely the motion along the rows of eyelets of the corresponding segments of the lacing pattern.

Because reflection of a segment does not alter its length, this representation leads to a path that has exactly the same length as the corresponding lacing pattern. The added advantage, however, is that it is now easy to compare the American and European patterns. In a few places they coincide, but everywhere else the American pattern runs along one edge of a thin triangle (one such triangle is shown shaded), while the European one runs along two edges of the same triangle. Because any two sides of a triangle exceed the third side in length (that is, a straight line is the shorter path between two given points), the American lacing is obviously shorter.

It is not quite so obvious that the shoe-store lacing is longer than the European. The simplest way to see this is to eliminate from both paths all vertical segments (which contribute the same amount to both lengths because each path has $n-1$ vertical segments) and also any sloping segments that match up. The result is shown in Figure 66 (dark lines). If each V-shaped path is now straightened out by reflection about a *vertical* axis placed at the tip of the V (faint lines), it finally becomes easy to see that the shoe-store path is longer, again because two sides of a triangle together exceed the third side.

For the shoelace problem, this cunning combination of graphical representations and reflection tricks can do more than just compare particular lacing patterns. Halton uses it to demonstrate that the American zigzag lacing is the shortest among *all* possible lacing patterns (the proof can be

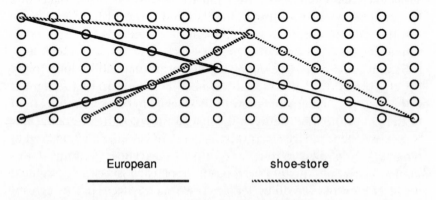

European   shoe-store

Figure 66 Eliminating common segments and comparing European and shoe-store lacing.

found in his article). More generally, both shoelaces and Fermat-style optics become united in the mathematical theory of geodesics – shortest paths in various geometries. There the reflection trick comes into its own in a big way, and Alice's mirror world sheds light on fundamental questions in physics, as well as confirming the superiority of the American way of lacing shoes.

STEINER NETWORKS

Shoe lacing is all very well, but manufacturers have a lot more on their mind than just the minimal length of lace. Style and fashion, for a start. And the shortest lacing for any given shoe design can be found by trial and error if you really need to. So my next, more ambitious application of optimisation is to communication networks, where practical solutions are harder to come by.

Peak District Cable, a new company with big ambitions but a small budget, wishes to lay a network of cables that will connect together the three British towns of Loughborough, Stoke-on-Trent, and Rotherham. It has two senior managers, Miles Spanning and Horatio Steiner. Spanning argues that since straight lines are the shortest paths between points, the best that can be done is to pick one town and link it to the others by two straight cables. Which town is picked depends on the actual distances, but it's simple enough to decide: look at the triangle formed by the towns and discard its longest side, then lay cable along the other two. Steiner concedes most of this, but has got it into this head that adding an extra town to the network might actually make it *shorter*. It may seem unlikely that he could be right – surely extra towns need extra cable – but if he *is*, where should the extra town be located to save the most cable, and how much does it save?

As it happens, Rotherham, Loughborough, and Stoke-on-Trent are all the same distance apart, 75 km. Spanning's network will be 150 km long, no matter which of the three towns is used as the central link. What about Steiner's hare-brained suggestion? Nestling in the Derbyshire dales near Matlock, roughly in the middle of the triangle formed by the three towns, is a village called – coincidentally – Middleton. It is roughly 44 km from each town. Steiner points out that if the centre of the network is located at Middleton, and one link is run from there to each of the three towns, then the total will be $3 \times 44 = 132$ km, a saving of 18 km, or about 12% (Figure 67).

The same questions can be asked for an arbitrary collection of towns.

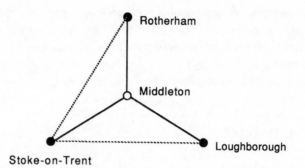

Figure 67   How to save cable by adding a town. The dotted network is 150 km long, the solid one only 132 km.

You can use Spanning's approach, and find the shortest network that joins them with straight links *without* introducing extra towns; or you can follow Steiner and invent new towns, which may shorten the network if you put them in the right place. In 1968, Edgar Gilbert and Henry Pollak, of AT&T's Bell Laboratories, conjectured that no matter how the towns are initially located, the maximum saving in cable that can be obtained by adding extra towns is 13.34%. Following the usual rules for mathematical attribution (name the idea after someone vaguely connected with the problem), this has become known as the *Steiner ratio conjecture*. After twenty-three years of unrelenting but largely unrewarded effort, it was finally proved by Ding Zhu Du at Princeton University and Frank Hwang at Bell Labs.

In its mathematical formulation, the towns to be connected are represented by points in the plane, and the cables linking them are straight lines. Whether or not we invent new towns, it's clear that the links must form a *tree* – the name for a network without any loops. Loops just waste cable, joining up towns that are *already* joined by some other route.

If no new towns are involved, then such a network is called a *spanning tree*. There are a lot of spanning trees to choose from, but in principle you can just list them all and see which is the shortest. For example, suppose there are four towns: Aylesbury, Brighton, Clacton, and Dagenham. Figure 68 shows some of the possible spanning trees and their lengths. The shortest one has a 'branch point' at Dagenham, from which three links run to the other three towns. On the other hand, if the towns are Ashburton, Bristol, Cheltenham, and Daventry, which are roughly in a straight line, then you can easily convince yourself that the shortest spanning tree joins them in that order and has no 'branch points'.

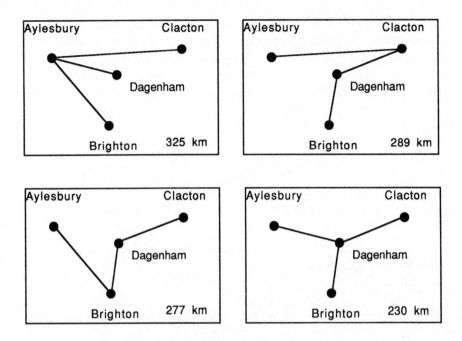

Figure 68   Four of the sixteen possible spanning trees for four English towns. The one at lower right is the shortest among all sixteen.

The problem is far more subtle if links are allowed to meet out in the countryside as well. For example, if there are three towns at the vertices of an equilateral triangle, as in our opening example (Rotherham, Loughborough, Stoke-on-Trent), then the shortest network joins all three to the centre of the triangle (Middleton – or in reality a nondescript place somewhere in a field not far from it). The shortest such network must also be a tree: it is called a *Steiner tree*.

What does Jacob Steiner have to do with the problem? As it happens, not much. He solved the problem for three towns in 1837. His name was attached to the problem by Richard Courant and Herbert Robbins in 1941, in their classic popularisation *What is Mathematics?* They, and Steiner, seem not to have known that he was beaten to the punch by Evangelista Torricelli and Francesco Cavalieri around 1640. They broke the problem down into two different cases. If the triangle has an angle of 120° or more, then the shortest network consists of just two links, joining that vertex to the other two. But if the triangle formed by the three towns has all its angles less than 120°, then the shortest network consists of three links that lead from the towns to the *Steiner point*, the unique place where all three roads meet at angles of 120º (Figure 69). Soap bubbles 'know' this. If you model the towns by three short pins, sandwiched between two sheets of Perspex, and then dip the whole thing into soap solution, then the soap forms three ribbons which meet at the Steiner point☞. A complete proof requires some sophisticated geometry, and here we'll take it for granted.

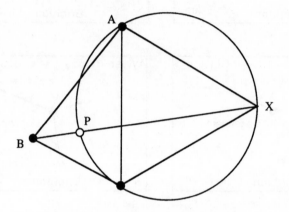

Figure 69   Constructing the Steiner point of triangle ABC: draw an equilateral triangle ACX; its circumcircle cuts BX at the Steiner point P.

Steiner also proved that when there are several towns, the edges of any Steiner tree must meet at 120º at each new town added, a simple consequence of the solution for three towns. He was less forthcoming on how to find such trees.

The problem of using a Steiner tree to join a larger number of towns than three was first investigated seriously by Milos Kössler and Vojtech Jarník in 1934. Finding the shortest Steiner tree in any given example requires a much more complicated calculation than finding the shortest spanning tree, because ever such a lot of new Steiner points have to be considered. For example, suppose there are six towns arranged at the corners of two adjacent squares, as in Figure 70. One possible Steiner tree is shown in Figure 70(a): it is found by solving the problem for a square of four towns first, and then linking in the two remaining towns via their Steiner point with one that is already hooked in. However, the shortest Steiner tree is that shown in Figure 70(b). You can't build up shortest Steiner trees piecemeal. Gilbert and Pollak asked whether the two versions of the problem might be related. Call the length of the shortest

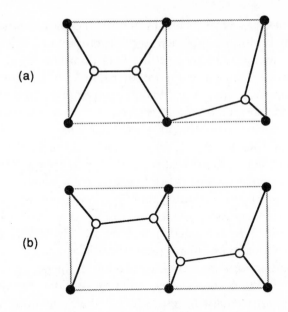

Figure 70 (a) Combining Steiner trees for a square and an isosceles right triangle; (b) shorter Steiner tree for the same set of towns.

spanning tree the *spanning length* of the set of towns, and that of the shortest Steiner tree *Steiner length*. Now, every spanning tree is also a Steiner tree (either invent no new towns, or put them on top of the existing links). So, for any set of towns, the spanning length is always greater than or equal to the Steiner length. How much greater can it be?

For an equilateral triangle of unit side, the spanning length is 2 and the Steiner length is $\sqrt{3}$. In this case, the ratio between the Steiner length and the spanning length is $\sqrt{3}/2=0.866$, and the saving in length obtained by using the shortest Steiner tree rather than the shortest spanning tree is about 13.34%. Gilbert and Pollak's Steiner ratio conjecture states that you can never do better than this. For any number of towns, arranged in any possible manner, the ratio of the Steiner length to the spanning length is always greater than or equal to $\sqrt{3}/2$. The saving in length obtained by using the shortest Steiner tree instead of the shortest spanning tree is never more than 13.34%.

ROUTING PROBLEMS

Similar (but in detail more complicated) problems arise in the routing of telephone lines, gas pipes, cable TV networks, buses, trains, and aeroplanes – and also in one approach to the evolution of living organisms. The genetic material of living creatures is DNA, which 'enodes' developmental information as a sequence of four types of base (adenine, thymine, cytosine, guanine). In this (grossly simplified) picture, genetic information is specified by a long sequence of chemical 'bases'. In the application of Steiner trees, the 'towns' are sequences of DNA in different organisms, and the 'distance' is some measure of the similarity between different sequences, such as what proportion of corresponding bases are equal. Steiner points correspond to 'most plausible common ancestors'. There is of course no guarantee that this common ancestor actually existed; but the method provides interesting clues to how the DNA molecule might have changed and how the organisms are related genetically. There are also many problems analogous to that of Steiner trees. The most important practical ones occur in the design of electronic circuits. Here the connections are generally laid out on a rectangular grid, running only horizontally or vertically, but the same kinds of question can be asked, and similar methods may help with their solutions.

The Steiner ratio conjecture is important for the 'economics' of all such networks, because the shortest spanning tree is much easier to find than the shortest Steiner tree, and it may therefore be worth sacrificing the 13.34% error to save computational effort. Indeed, much of the work on the Steiner ratio conjecture has emerged from Bell Labs, because AT&T is a telephone company. Until quite recently, for convenience, it used the spanning length to charge customers who wanted to connect their offices together. It was worried that customers might discover that they could make major savings on their bills by inventing imaginary offices located at appropriate Steiner points! The conjecture limits such savings to 13.34%, which is not too embarrassing. Why didn't AT&T save itself the worry and use the Steiner length itself?

It couldn't.

Finding the spanning length is a simple computation, even for a huge number of towns. It is solved by the *greedy algorithm*: start with the shortest link you can find, and at each stage thereafter add on the shortest remaining link that doesn't complete a closed loop, until every town is included in the tree.

Finding the shortest Steiner tree isn't so easy. You can't just do it by taking all possible triples of towns, finding their Steiner points, and then looking for the shortest tree that joins the towns together and meets either at towns or at Steiner points. The correct generalisation of 'Steiner

point' to a set of many towns is
any point at which links can meet
at 120°. For as simple an example
as four towns at the vertices of a
square, these points are not Steiner
points of any subset of three towns
(Figure 71). There are infinitely
many points in the plane, and
even though most of them are
probably irrelevant, it is not
obvious that finite algorithms exist
at all! In fact they do; the first was
invented by Z.A. Melzak of the
University of British Columbia, but
his method becomes unwieldy
even for moderate numbers of
towns. It has since been improved,
but not dramatically.

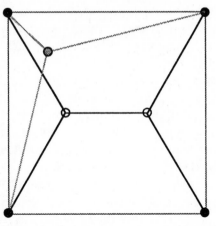

Figure 71   Steiner points (white) for
four towns in a square (black) are different from the Steiner point of a subset of
three towns (grey).

ALGORITHMIC COMPLEXITY

We now know that there are good reasons why these solutions are
inefficient. The growing use of computers has led to the development of
a new branch of mathematics: complexity theory. This studies not just
algorithms – methods for solving problems – but how efficient those
algorithms are. Given a problem involving some number n of objects
(here towns), how fast does the running time of the solution grow as n
grows? If the running time grows no faster than a constant multiple of a
fixed power of n, such as $5n^2$ or $1066n^4$, then the algorithm is said to run
in *polynomial time*, and the problem is considered to be 'easy'. Usually this
means that the algorithm is practical (but it will not be if the constant is
absolutely huge). If the running time grows non-polynomially – faster
than any constant multiple of powers of n, for instance exponentially,
like 2^n or 10^n – then the problem has non-polynomial running time and
is 'hard'. Usually this means that the algorithm is totally impractical. In
between polynomial time and exponential time is a wilderness of 'fairly
easy' or 'moderately hard' problems, where practicality is more a matter
of experience.

For instance, adding two n-digit numbers requires at most $2n$ one-digit additions, including carries, so the time taken is bounded by a constant multiple (namely 2) of the first power of n. Long multiplication of two such numbers involves about n^2 one-digit multiplications and no more than $2n^2$ additions, or $3n^2$ operations on digits, so now the bound involves only the second power of n. The opinions of schoolchildren notwithstanding, these problems are therefore 'easy'. In contrast, consider the travelling salesman problem: find the shortest route that takes a salesman through a given set of cities. If there are n cities, then the number of routes we have to consider grows faster than any power of n. So case-by-case enumeration is hopelessly inefficient.

Oddly enough, the big problem in complexity theory is to prove that the subject actually exists – that is, to prove that some 'interesting' problem really is hard. The difficulty is that it is easy to prove that a problem is easy, but hard to prove that it is hard! To show that a problem is easy, you just exhibit *one* algorithm that solves it in polynomial time. It doesn't have to be the best, or the cleverest: any will do. But to prove that a problem is hard, it is not enough to exhibit some algorithm with non-polynomial running time. Maybe you chose a silly one, maybe there's a better one which *does* run in polynomial time. You have to find some mathematical way to consider *all possible algorithms* for the problem, and show that *none* of them runs in polynomial time.

There are lots of candidates for hard problems – the travelling salesman problem, the bin-packing problem (how can you best fit a set of items of given sizes into a set of sacks of given sizes?), and the knapsack problem (given a fixed size of sack, and lots of objects, does any set of objects fill the bag exactly?). So far nobody's managed to prove that any of them are hard☛. However, in 1971 Stephen Cook of the University of Toronto showed that if you can prove that any one problem in this candidate group really is hard, then they all are. Roughly speaking, you can 'code' any one of them to become a special case of one of the others: they sink or swim together. These problems are called *NP-complete*. Everyone *believes* they really are hard, if only on the psychological grounds that you can't expect to get all good things at once. Ron Graham, Michael Garey, and David Johnson of AT&T have proved that the problem of the Steiner length is NP-complete. An efficient algorithm to find the precise Steiner length for any set of towns would automatically lead to efficient solutions to all sorts of computational problems that are widely believed not to possess such solutions. The Steiner ratio conjecture is therefore important, because it proves you can replace a hard problem by an easy one without losing very much.

LATTICE TREES

The equilateral triangle example that started the whole business is very natural. It suggests that there must be a simple proof of the conjecture. However, its simplicity may be deceptive, for if there is a simple proof, nobody has ever found it. Direct attacks for small numbers of towns lead to vast and messy calculations. Gilbert and Pollak had quite a lot of evidence for their conjecture; and in particular they could show that something along those lines must be true: they proved that the ratio is always at least 0.5. By 1990, various people had performed heroic calculations to verify the conjecture completely for networks of four, five, and six towns. For general arrangements of as many towns as you like, they also pushed up the limits on the ratio from 0.5 to 0.57, 0.74, and 0.8. Not long ago, Ron Graham and Fang Chung at Bell Communications Research raised it to 0.824, in a computation they describe as 'really horrible – it was clear it was the wrong approach'.

To make further progress possible, the horrible calculations had to be simplified. Du and Hwang found an approach that is so much better – it does away with the horrible calculations completely. The basic question is how to get equilateral triangles in on the act. There's a big gap between the triangle example, which sets up the bound on the ratio, and a general system of towns, which is supposed to obey the same bound. How can this no man's land be crossed? There's a kind of halfway house. Imagine the plane tiled with identical equilateral triangles, in a triangular lattice (Figure 72). Put towns only at the corners of the tiles. It turns out that the only Steiner points that need to be considered are the centres of the tiles. In short, you have a lot of control, not just on computations, but on theoretical analyses.

Of course, not every set of towns conveniently lies on a triangular

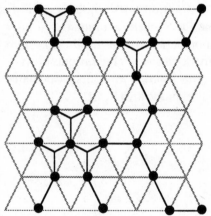

Figure 72 A Steiner tree for towns that lie on a triangular lattice has a much more rigid and regular structure than that for general towns. Du and Hwang reduce the Steiner ratio conjecture to the same problem for lattice trees.

lattice. Du and Hwang's insight is that the crucial ones do. Suppose the conjecture is false. Then there must exist a *counter-example*: some set of towns for which the ratio is *less* than $\sqrt{3}/2$. They show that if a counter-example to the conjecture exists, then there must be one for which all the towns lie on a triangular lattice. This introduces an element of regularity into the problem, and it is then relatively simple to polish it off.

How to prove this lattice property? It's a wonderful exercise in lateral thinking. First they reformulate the conjecture as a problem in game theory, where players compete and try to limit the gains (payoff) made by their opponents. Game theory was invented by John Von Neumann and Oskar Morgenstern in their classic *Theory of Games and Economic Behavior* of 1947. In the Du–Hwang version of the Steiner ratio conjecture, one player selects the general 'shape' of the Steiner tree, and the other picks the shortest one of that shape that they can find. Du and Hwang deduce the existence of a lattice counter-example by observing that the payoff for their game has a special 'convexity' property.

This elegant new method neatly disposes of a question that previously looked totally intractable, and cuts a broad swathe through the distinctly *non*-magical maze of tangled calculations and case-by-case investigations. The method does require a certain amount of mathematical technique; but it is such a dramatic improvement that it knocks all previous approaches on the head. More importantly, it provides a paradigm for investigating analogous questions.

The motto 'think first, calculate later' should be engraved on every mathematician's heart.

JUNCTION EIGHT

*T*he entrance to the eighth passage of the magical maze is easy enough to find – a tall, narrow wooden gateway between two huge pillars. It looks for all the world like the entrance to Jurassic Park, in the movie of the same name.

Maybe it is. Written on it, in huge red letters, is the phrase BEWARE OF MONSTERS. *Scrawled just below, however, in what looks like felt tip, is a graffito:* if you're a monster, beware of meteorites.

Plants run riot around and over the gateway – not flowering plants, like those we encountered at the entrance to the magical maze, but ferns.

Huge, prehistoric *ferns.*

A fern frond rustles. You step back, expecting a Tyrannosaurus rex *to rush out. Or maybe a velociraptor. Instead, it's a rabbit. A white rabbit, for all the world just like the pair you saw what seems an age ago at the maze's entrance. To your relief, it doesn't carry a pocket watch. But it does carry a felt tip pen.*

A second rabbit emerges from behind the ferns; this one is black. Then another, and another. A whole family of black rabbits troops to the side of the gate, and disappears behind it through a tiny door in the front of the pillar.

The last rabbit to pass through is the white one. It turns, and beckons you to follow. You approach the gate, ready to push it open, but it swings ponderously aside on squeaky hinges.

Beyond is a narrow pathway, paved in bricks.

The pattern of the bricks is funny. They are square slabs, arranged around a series of square holes, each hole the same size as a single slab. A fern is planted in each hole. The slabs have square holes in their centres, too – and a smaller fern sprouts from each. And as you look more closely, you see more and more ever-tinier square holes, and more and more ever-tinier ferns.

The bricks, of course, are yellow.

The wall is a mosaic of triangular tiles. Again, the tiles are arranged to leave holes, and what at first sight looks like a single tile turns out, on close inspection, to be three smaller triangular tiles surrounding an upside-down triangular hole.

On shelves, along both walls, are elaborate pottery and metal boxes, each containing a rock. Tucked neatly into the corner of the boxes are strange instruments – tiny magnifying glasses, mirrors, fans, watering-cans, miniature glass bottles ...

You pick up a bottle. With the aid of one of the magnifying glasses, you decipher the writing on its label: 'Acid rain, bottled in 1983'. You put the bottle back on the stand, and as you do so you spot a battered piece of card lying on the floor. When you turn it over it bears the message BONSAI MOUNTAINS – £50–£5000.

Also written in felt tip.

The white rabbit pops its head round the distant corner, and once more beckons you to follow. The only alternative is to go back; what have you got to lose? You swallow hard, try not to think about monsters – or meteorites – and set off along the yellow brick road.

Passage Eight

GALLERY OF MONSTERS

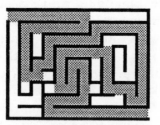

The eternal battle between order and disorder runs like a deep ocean current through the human perception of the universe, a common feature of many creation myths from many cultures. In the Old Testament, 'the Earth was without form, and void, and darkness was upon the face of the deep'. In the Enuma Elish, an early Babylonian epic, the universe arises from the chaos that ensues when an unruly family of gods is destroyed by its own father. Order is equated with good and disorder with evil. Order and chaos have always been seen as two opposites, twin poles about which we pivot our understanding of the world.

Not any longer. Today, at the frontiers of humanity's exploration of the magical maze, we are discovering that order and chaos are not opposites, but soulmates – two sides of the same coin, two edges of the same sword. A popular version of this new viewpoint has leapfrogged the tiresome process of objective scientific testing, establishing itself as 'chaos theory' – a world of psychedelic posters and postcards, anarchistic philosophies, maverick guru scientists, and devastating hurricanes caused by one flap of a butterfly's gossamer wing.

The media have attributed almost mystical powers to chaos theory.

When they're not rubbishing it, that is.

There is a mathematician in *Jurassic Park,* called Ian Malcolm, and he is a chaos theorist. He knows that Jurassic Park's complex systems are doomed, right from the start, but nobody listens until it's too late. This is chaos theory as Cassandra. In fact, the movie doesn't actually tell us much about chaos, and most of what it does say misses the point, which is about what you'd expect from Hollywood. The book version is better, and it avoids the trite 'Man was not intended to meddle' ending. Not that the book tells you much about chaos theory either.

So I guess it's up to me☛.

Chaos theory has two main ingredients: geometrical shapes known as 'fractals', and irregular behaviour called 'chaos'. The two concepts evolved separately, but have since become inseparably intertwined. Chaos theory sheds new light on the predictability of nature – and also muddies the waters. Cassandra it is not, but it sometimes sounds just like her. Chaos theory provides a new angle on an old discovery: 'laws of nature'. The revolution of scientific thought that culminated in Isaac Newton led to a vision of the universe as some gigantic mechanism, functioning 'like clockwork', slavishly obeying simple, fixed mathematical principles. Philosophers call this idea 'determinism'. In particular, it led Pierre Simon de Laplace – one of the great eighteenth-century mathematicians – to an astonishing vision of a vast intellect, capable of predicting the behaviour of the entire universe from a single formula. In a fully deterministic universe, Laplace pointed out,

> an intellect which at any given moment knew all the forces that animate nature and the mutual positions of the beings that comprise it, if this intellect were vast enough to submit its data to analysis, it could condense into a single formula the movement of the greatest bodies of the universe and that of the lightest atom: for such an intellect nothing could be uncertain, and the future, just like the past, would be present before its eyes.

In Douglas Adams' *The Hitch Hiker's Guide to the Galaxy*, Laplace's vision was parodied in the ultimate supercomputer Deep Thought, with its answer 'forty-two' to the great question of Life, the Universe, and Everything. In Laplace's disembodied super-intellect we find the paradigm of the clockwork universe, never deviating from its initial course once its cogwheels have been set in motion. And, for all its faults, it has been spectacularly successful in helping humanity to come to terms with the world around it.

But now, mathematicians and physicists have discovered something rather curious. Order can breed its own peculiar kind of disorder. Deterministic causes – equations that do not contain any random terms, which in principle describe the evolution of some system uniquely for all time –

can have random effects.

This is a remarkable discovery, and it is changing the face of science. It is known as deterministic chaos, or just plain chaos. Chaos lies at the frontiers of today's mathematics, one of several startling new paradoxes about the way the world can change. Others include the self-organisation of evolutionary systems, their self-complication, the spontaneous emergence of 'computational' entities, and the emergence of large-scale order from small-scale disorder. Nature is far more complicated, far more interesting – and far more clever – than we think. Its patterns are not the direct consequences of simple laws, but emerge indirectly from an all-embracing sea of chaos and complexity.

Take your own heart. Traditional science treats it as a pump, beating 'like clockwork', whose cycles can be dissected into simple waves of standard shapes. Real hearts are far more puzzling. Your heartbeat is triggered by signals from your brain, but the rhythmic contractions that keep you alive are the result of a democratic vote by millions of muscle fibres, all agreeing to contract in synchrony. Like the flashing fireflies.

A simple, clockwork process?

Even when your body is at rest, your heartbeat varies by tiny but measurable amounts. This variability is not the result of random outside influences: it is caused by chaotic internal dynamics. There are good reasons for the chaos: your heart wouldn't work without it. Chaos distributes wear and tear more evenly; it makes your heart able to respond more rapidly to changes in your surroundings. A clockwork heart would only work in a clockwork person.

THE CHAOS GAME

A parallel paradox is that genuinely random processes can lead to a surprising degree of order. In his book *Fractals Everywhere*, the British-born mathematician and entrepreneur Michael Barnsley introduces what he – rather confusingly – calls the Chaos Game.

Really, it should be called the Fractal Game – but who am I to contradict its inventor? After all, look what happened to the people who disagreed with the mathematician in *Jurassic Park*.

The Chaos Game is both simple and surprising. In a daring act of imagination, Barnsley has parlayed its simplicity, and its surprise, into a multi-million dollar industry. Behind the Chaos Game lie ideas that have transformed the way we store and transmit visual images. Related ideas have given mathematicians a completely new way to model nature, one whose implications are only just beginning to be recognised.

To play the Chaos Game, you need a piece of paper, a ruler, a pencil, and a 'three-sided coin' for which 'heads', 'tails', and 'edge' have the same probability, 1/3. For example, you could use a die, and let 1 or 2 = heads, 3 or 4 = tails, and 5 or 6 = edge. Mark three points on the sheet of paper, say at the vertices of an equilateral triangle. Label the three corners of the triangle 'heads', 'tails', and 'edge'. Use your pencil to mark a random point on the paper. Toss the 'coin', and move the point half-way towards the corresponding vertex, getting a new point. Draw that, too. Repeat this procedure, starting from the new point, and always generating the next point from the previous one by tossing the 'coin' and moving half-way towards the appropriate vertex (Figure 73).

What do you see?

You might expect the result to be some uniform cloud of points in the plane. Not so. Figure 74 shows what happens with a thousand trials, drawn using a computer.

This strange shape is the *Sierpiński gasket*, and we've already encountered something very like it in the chapter 'Panthers Don't Like Porridge', in connection with the Tower of Hanoi. The gasket was invented by the Polish mathematician Wacław Sierpiński: for him it was an example of a curve that crosses itself at every point. The gasket is constructed from an equilateral triangle. Divide it into four equal quarters, and throw away the middle one. This leaves three triangles half the size. Divide each into four equal quarters, and throw away the middle one. This leaves nine triangles one-quarter the original size. Repeat for ever: the gasket is what remains when everything else has been thrown away. Figure 75 shows what it looks like after the first five stages of its construction.

The Sierpiński gasket seems a very odd shape to be generated by a random procedure. In fact, it seems a very odd shape – period. When such shapes first appeared on the mathematical scene, about a century ago, they were derided as a 'gallery of monsters'. At that time their main purpose was to shine a light into the murkier corners of the magical maze by showing how nasty mathematics could get. Several leading mathematicians saw little point in doing that, and said so – often with some sarcasm.

Today, however, we recognise that shapes such as Sierpiński's are entirely natural – and useful. His gasket is one of the more regular representatives of a class of mathematical shapes called *fractals*. Fractals

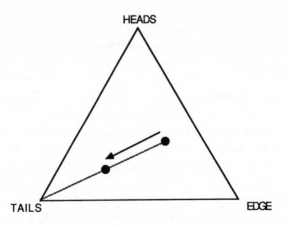

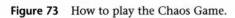

Figure 73   How to play the Chaos Game.

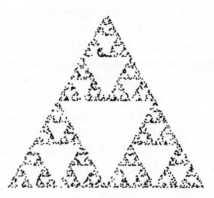

Figure 74   A thousand trials of the Chaos Game. Familiar?

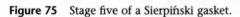

Figure 75   Stage five of a Sierpiński gasket.

were invented, named, and promoted by Benoît Mandelbrot, and they are a new way of modelling the irregularities of nature.

BONSAI MOUNTAINS

The shapes studied in classical geometry are things like triangles, squares, lines, circles, ellipses, spheres, and so on. They are all very simple shapes, and they share a common feature: they have no interesting structure on sufficiently small scales.

Consider a circle, for example. Imagine looking at it through an enormously powerful microscope. In the field of vision of the microscope you see a tiny portion of the circle, magnified to an immense size. When the magnification is low, you see a curved circular arc. As the power is turned up, you still see a circular arc, but at high scales of magnification it no longer looks curved. It is curved, but the curvature is so slight that the eye no longer notices it. At high magnification, a circle looks almost exactly like a straight line, with no interesting features whatsoever.

The same goes for a sphere: at high magnification, a sphere looks almost exactly like a plane, again with no interesting features whatsoever. This is why primitive cultures thought the Earth was flat. It *looks* flat when all you can see is a few square kilometres of its vast, gently rounded surface.

When you magnify a Sierpiński gasket, however, it does not become featureless. Inside the gasket are triangular holes, as tiny as you wish. However much you magnify the gasket, the holes will still be there. Holes so tiny that your eye cannot see them will grow until they almost fill the field of vision. In fact, no matter how much you magnify a Sierpiński gasket, what you see will always look much like a Sierpiński gasket. Unlike the sphere and circle, the Sierpiński gasket has fine detail at any scale of magnification.

In fact, the Sierpiński gasket is built from three identical copies, each one half the size (Figure 76). Each of those is again made from three identical copies, each one half the size – so the Sierpiński gasket is built from nine identical copies, each one-quarter the size, too. By repeating this argument we see that it can be made from 27 identical copies, each 1/8 the size; 81 identical copies, each 1/16 the size; 243 identical copies, each 1/32 the size ... and so on indefinitely. This proves that we can never run out of detail, because however small the piece you look at, it will contain a tiny, perfect copy of the entire gasket. We say that the gasket is *self-similar*: tiny bits of it look much like the entire object.

Figure 76   A Sierpiński gasket is made from three perfect copies, each half as big.

Many shapes in nature have the same kind of delicate, hidden, self-similar detail. Unlike a true mathematical fractal, whose detail goes on for ever, real shapes do eventually fuzz out when we get down to atomic scales – but, even so, they are much better modelled by a fractal than they are by a sphere or a circle.

Mountain ranges have structure on very small scales. A mountain range is a collection of jagged peaks. Each peak is itself a collection of jagged sub-peaks, and so on. A lump of rock broken off a mountain looks very much like a miniature mountain: this is why Terry Pratchett's book *Small Gods* can feature a character whose hobby is 'bonsai mountains'. You'll recall that the Japanese art of bonsai produces miniature trees by growing small shrubs and training them into the shape of a mature tree. The process takes years, and uses all sorts of special equipment and techniques. Bonsai is possible because a small part of a tree has a very similar structure to the whole tree. Pratchett's six-thousand-year-old 'history monk' Lu-Tze can culture bonsai mountains – using equally specialised equipment, such as tiny mirrors to simulate the glare of the sun and a watering-can to simulate the erosion caused by thunderstorms – because a jagged lump of rock retains all of the complexity of a huge mountain range. After all that practice, he does it extremely well. As the novice Brutha asks, 'That can't really be snow on the top, can—' at which point he is interrupted by the Great God Om, who is a small tortoise.

Coastlines have the same property as bonsai mountains. If you look at a map of a county, country, or continent that possesses a coast, you'll find that the coastline always looks much the same – not in detail but in 'texture' – whatever the scale of the map. Maps drawn to a larger scale show more detail, but the general nature of that detail is always the same: bays, promontories, random-looking wiggles. Coastlines are 'statistically self-similar': little bits of them look like big bits of some *other* possible coastline.

Many plants behave in the same manner, too. I've already mentioned trees. Ferns are beautifully fractal: a fern leaf is made from a series of fronds, sticking out to the left and right from a central spine. Each frond is made from a series of sub-fronds, each sub-frond from a series of sub-sub-fronds, and so on – for four of five steps, typically. Even more striking is broccoli romanesco (Figure 77), a cauliflower-like plant. Its head consists of a spiral swirl of florets. Each floret is a spiral swirl of sub-florets, each sub-floret is a spiral swirl of sub-sub-florets ... and so on.

Bonsai coastlines, bonsai ferns, bonsai broccoli. All of them are far better modelled by fractals than by sphere, cones, cubes, pyramids, or the other paraphenalia of Euclid's geometry.

Figure 77   The fractal self-similarity of broccoli romanesco.

FRACTAL DIMENSION

In order to be able to do science using fractal models, we need a way to characterise a fractal quantitatively. We need a number that can be measured in an experiment, and captures some useful flavour of the fractal's geometry. The most important such quantity is known as the fractal dimension.

We'll sneak up on fractal dimension by thinking about more traditional shapes – lines, squares, cubes. In ordinary mathematical parlance, a line is 1-dimensional, a square is 2-dimensional, and a cube is 3-dimensional. One way to see this is to notice that a line pokes out along one direction (say east), a square pokes out along two independent directions (east and north), and a cube pokes out along three directions (east, north, up). But there's different way to work out the dimension, which uses scaling.

Suppose you want to make a line twice as big. You can do so by joining together two copies (Figure 78(a)). If you want to make the line three times as long, you can do so by joining together three copies (Figure 78(b)). And so on. To make the line n times as long, you join together n copies.

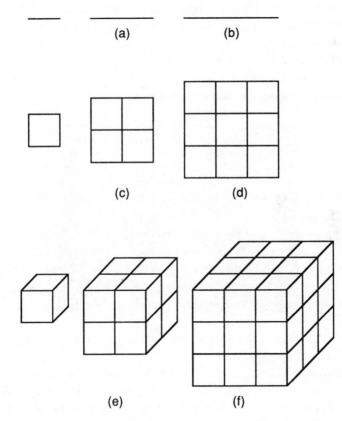

Figure 78   The effect of dimension on scaling: (a) making a line twice as big, or (b) three times as big; (c) making a square twice as big, or (d) three times as big; (e) making a cube twice as big, or (f) three times as big.

Next, suppose you want to make a square twice as big. Then you can do so by joining together *four* copies (Figure 78(c)). If you want to make the square three times as big, you can do so by joining together nine copies (Figure 78(d)). To make the square n times as big, you join together n^2 copies.

Continuing up the dimensions, suppose you want to make a cube twice as big. Then you can do so by joining together *eight* copies (Figure 78(e)). If you want to make the cube three times as big, you can do so by joining together twenty-seven copies (Figure 78(f)). To make the cube n times as big, you join together n^3 copies.

A square is the 2-dimensional analogue of a cube; a line segment is the 1-dimensional analogue. Suppose we choose a dimension d. Then we can summarise our findings on the scaling of d-dimensional 'cubes' thus: to make a d-dimensional 'cube' n times bigger, you need n^d copies.

We can 'solve' this statement for d, if we're prepared to use logarithms☞. Let

$$k = n^d$$

be the number of copies. Taking logarithms, we get

$$\log k = d \log n,$$

so

$$d = (\log k)/(\log n).$$

The dimension of a 'cube' is the logarithm of the number of copies, divided by the logarithm of the size.

Suppose we apply all this to the Sierpiński gasket, without worrying too much (yet) about what the calculation signifies. We saw that the Sierpiński gasket can be doubled in size by taking three identical copies. So $n = 2$, $k = 3$. Plugging those numbers into the formula for d, we find that the dimension of the Sierpiński gasket is

$$d = (\log 3)/(\log 2),$$

which is about 1.584962.

The dimension of the Sierpiński gasket is not a whole number!

Well, that certainly explains why it looks so strange. 'Gallery of monsters', indeed.

You could choose to dismiss the calculation as meaningless: how can a Sierpiński gasket 'poke out along 1.58 directions'? But the imaginative mathematician follows his or her nose when they blunder into a new corner of the magical maze. And the trained mathematical nose, in this instance, smells something interesting. For what we are witnessing is this: the traditional concept of dimension can be extended to fractals, provided we focus on scaling properties, and *not* on 'number of directions'. The former generalises to fractals, the latter does not.

We say that the Sierpiński gasket has *fractal dimension* 1.58. The interpretation is that it 'scales' in a manner that lies somewhere between what a 1-dimensional object would do and what a 2-dimensional object would do. It takes two copies of a line to double its size, and four copies of a square to double its size. In between is the Sierpiński gasket, for which only three copies are needed. So it's entirely reasonable that its 'dimension' should lie between 1 (the dimension of the line) and 2 (the dimension of the square).

Figure 79 shows another fractal, the Sierpiński carpet. Here a square is repeatedly divided into nine sub-squares, each one-third the size, and the central sub-square is removed. What is the fractal dimension of the Sierpiński carpet? Well, it takes $k = 8$ copies to make the size increase by a factor of $n = 3$. So the fractal dimension is

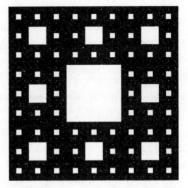

$$d = (\log 8)/(\log 3) = 1.892789.$$

Figure 79 The Sierpiński carpet.

Notice that this also lies between 1 and 2, but now it is a bit bigger than the dimension of the Sierpiński gasket. This is again reasonable, because the Sierpiński carpet is distinctly less 'holey' than the gasket. The fractal dimension captures 'how well the shape fills space', or perhaps 'how irregular it is'.

FAMILY TREES

Fractals can be used for all sorts of purposes. Scientists use them to model the growth of loosely knit clusters, such as soot; here computer models of the growth process generate fractals (Figure 80) whose dimension is very close to that measure in real soot. The scientist concludes that the model growth process bears some degree of resemblance to the real one, and so learns more about the formation of soot. Metallurgists use fractals to understand crystal growth in solidifying metals. Meteorologists use fractals to understand clouds. Very recently, a team of medical researchers discovered that they could explain some puzzling numerical features of blood flow if they assumed a fractal model for the body's veinous system.

Here, I want to show you a little of how fractals model plants. The idea is to set up simple rules that generate the branching pattern of the plant, and then represent the branches geometrically. The result is a self-similar branching struc- ture, a fractal plant. (It is self- similar in that any branch broken off obeys the same rules, and so is

Figure 80   Fractal model of soot.

a smaller version of the same 'species'.) The rules are called *Lindenmayer systems* after their inventor, Aristid Lindenmayer, or L-systems for short.

One of the simplest L-systems arises in connection with Fibonacci's rabbit problem. Remember how that goes. We start with a pair of immature rabbits. Each breeding season, the immature pairs mature for a season, while each mature one breeds a new immature pair. Earlier we asked how fast the population grows, but now we are going to look at the shape of the rabbits' family tree. 'Tree' starts out by being a metaphor here – and ends up a lot closer to reality. Fibonacci numbers don't just describe the spirals in sunflower heads, they are the visible tip of a marvellous mathematical theory of branching structures. The same scheme illuminates not just the numerology of plants, but their entire form – the way they, too, branch. We can now – in computer simulation – *grow* realistic grasses, flowers, bushes, and trees from mathematical rules. And there is a strong suspicion that those rules lie at the heart of how plants themselves grow.

We can summarise the growth pattern of the rabbits by using a 'grammar' in which the letters I and M stand for immature and mature pairs, respectively. Then the growth rules lead to a system of transform- ations, which take us from one season to the next:

> I→M   (immature pairs mature after one season)
>
> M→MI   (mature pairs stay alive and breed a new immature pair).

We start with one immature pair (I) and repeatedly apply the two branching rules to get the sequence of symbol strings

$$I \to M \to MI \to MIM \to MIMMI \to MIMMIMIM \to \ldots$$

At each step we apply the branching rules to every symbol in the string, and this leads to the next string. For example, to go from MIM to MIMMI we do this:

$$
\begin{array}{ccc}
M & I & M \\
\downarrow & \downarrow & \downarrow \\
MI & M & MI.
\end{array}
$$

Incidentally, we can solve the original Fibonacci problem 'how big is the rabbit population?' using this 'grammar', just by counting the symbols. For example, the final generation listed contains five M's and three I's, a total of eight pairs of rabbits. These are three consecutive Fibonacci numbers, and the pattern continues indefinitely.

Instead of counting the symbols, we can *interpret* them as branches in a tree diagram (Figure 81). Now we are modelling a plant which has two

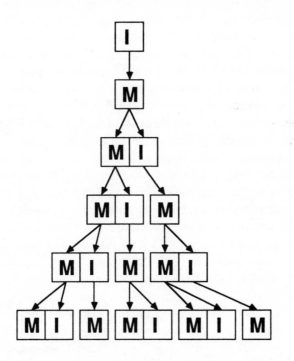

Figure 81   The diagram determined by Fibonacci's rabbits.

cell-types: immature ones, which mature for a season and then branch, and mature ones, which produce an immature side-branch while continuing to grow themselves. No longer a family tree of rabbits, this is just a tree – or some tree-like plant. If, say, the immature cells are round blobs and the mature ones are long and thin, we get a plant something like Figure 82.

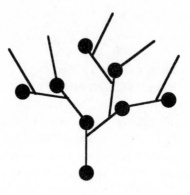

Figure 82   A plant designed using the rabbits' family tree.

L-systems help to explain the patterns in which plants branch. If you look closely at plants and shrubs, you will often find that their branches don't divide at random but have regularities. Perhaps branching will lead to one long stem and one short one, each of which branch in turn according to the same pattern, for instance. Figure 83 shows some plant-like shapes generated by various L-systems.

So, by a curious quirk of fortune, it turns out that branching patterns in plants, as well as numbers of petals, depend on mathematics that arose from Fibonacci's rabbit problem. But now it is the rabbit's family tree, not the size of their population, that matters.

MATHEMATICS CAN BE FERN

Barnsley's image processing methods also stemmed (pun intended) from a plant – the black spleenwort fern. This beautiful fractal (Figure 84) is made from four slightly distorted copies of itself. Musing on the how such remarkable complexity can be generated by simple fractal rules, Barnsley suddenly realised that the fern contained the germ of something

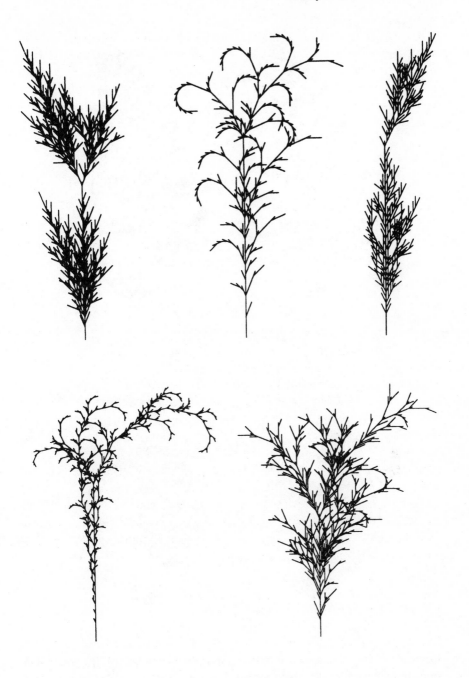

Figure 83   Plant shapes created by L-systems.

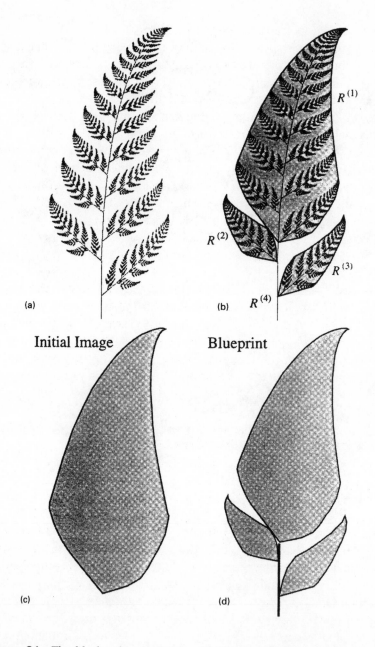

Figure 84 The black spleenwort fern. (a) The fern fractal. (b) The four transformations $R^{(1)}$, $R^{(2)}$, $R^{(3)}$, $R^{(4)}$, that generate it. $R^{(4)}$ squashes the whole fern flat on top of its stalk. (c) Initial course outline of fern. (d) The four transformations pictured as a geometric 'blueprint'.

potentially very important: image compression.

In today's world, we spend a lot of time and money to send images all over the globe. Television – terrestrial, satellite, cable – does little else, at 25 images per second. Many images are transmitted over the Internet. Businessmen fax pictures to their distant colleagues.

When you fax someone an image, the fax machine scans it, row by row, and turns the picture into a long series of binary signals – 0 for white, 1 for black, say. (The actual process is more sophisticated, but this description will do.) Then the signals are sent along a phone line as a series of bleeps. At the other end, the receiving fax decodes the bleeps and turns them back into black and white dots, which the human eye sees as a picture. TV works the same way, but with colour information included too.

One picture generates an awful lot of dots. At, say, 300 dots per inch, a picture the size of a postage stamp requires about 100,000 dots. A full-page image takes perhaps 10 million.

Suppose you want to send someone a full-page picture of the black spleenwort fern, by fax, using the conventional approach. Then you need to send 10 million binary digits. But – and this is the big insight – you can transmit the fractal *rules* for constructing the fern using only a few hundred binary digits – one-thousandth of 1% of the information. Of course, the person at the other end has to know how to turn those rules into a picture: this is where the Chaos Game comes in, but the details can wait for a few pages.

This reduction of information – the technical term is 'data compression' – also reduces transmission time, and hence cost. If you could compress video images by this amount, then every satellite TV channel could be replaced by 100,000 channels.

Of course, it's not *that* easy. TV involves sound as well as images. It takes time to turn an image into its fractal rules, and it takes more time to reverse that process at the other end. And, crucially, viewers want to look at things other than the black spleenwort fern. Soaps, sitcoms, blockbuster movies, Formula One, golf ... Fortunately, the compression procedure is quite general: it works on any fractal. The world, as we've seen, is full of fractals. Most soaps and sitcoms have plants in the picture. Blockbuster movies happen in jungles, on mountains, or on the cratered surfaces of alien planets. Formula One has big cars in the foreground – and smaller ones, of much the same shape, in the background, trying desperately to catch up with the leader. Golf has trees all over the place, often between the ball and the hole. So Barnsley's idea was a promising start, and further research would surely improve it.

Barnsley tried to sell his marvellous idea to various major communication companies. No luck. It was too radical. So he set up his own company. And, after a lot of effort and a great deal of worry, he discovered that a variation on his idea works for *any image whatsoever*. The degree of compression is no longer by a factor of 100,000, mind you, but compression by a factor of 100 or so – reducing the picture information to 1% of its original size – is entirely feasible.

The method now works like this. A computer scans the image looking for parts of it that resemble other parts, but on a larger scale. If the computer can assemble a long enough list of such related pairs of parts of the image, then the entire image can be reconstructed from that list. The information needed to transmit the list is far less than that needed for the image itself. Figure 85 shows an image compressed using Barnsley's technique. The Encarta™ CD-ROM encyclopaedia contains 8,000 full colour images, compressed using Barnsley's fractal methods. Without image compression, it would have been more like 80 images.

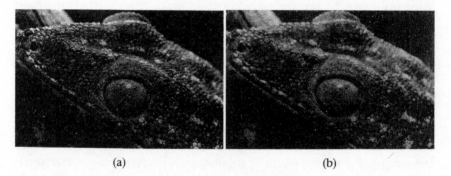

(a)   (b)

Figure 85 Image compression by Barnsley's method: (a) image of a gecko, uncompressed; (b) the same image compressed to require 1/156 as much memory.

How do we reconstruct an image from fractal rules?

Let's think about the fern. The rules boil down to a list of *transformations*, here four in number. Each transformation tells you how to take the plane, distort it, and map it on to itself. The fern is made up from four transformed copies of itself, using those four transformations.

It is, in fact, the *only* shape that is made up from four transformed copies of itself using those four transformations. In principle, then, knowing the transformations implies that you know the shape.

How, though, can you actually *find* the shape from the list of transformations?

Here's a really neat, clever method. Call the four transformations T_1, T_2, T_3, T_4, say. Imagine putting them into a bag. Choose some point P_0 in the plane, and draw a dot. Now begin pulling transformations out of the bag, at random.

Say you pull out T_3 first. Apply that transformation to P_0 to get a new point:

$$P_1 = T_3(P).$$

Draw that point in the plane too. Put T_3 back in the bag, and again pull out a transformation at random – say, T_4. Now apply that transformation to P_1 to get a new point:

$$P_2 = T_4(P_1).$$

Draw that point in the plane.

Continue this process, pulling out a transformation at random, applying it to the last point drawn to get a new point, drawing that point on the plane, and putting the transformation back in the bag. You slowly build up a fuzzy cloud of points. After a while, the cloud starts to look suspiciously like the black spleenwort fern. The longer you continue the process, the closer the resemblance gets.

There are other ways to reconstruct the fractal from its rules, but this method is quick, simple, and surprising.

What of the Chaos Game? That is the same method applied to the Sierpiński gasket. The gasket is formed from three copies of itself. To get these copies, choose one of the three corners of the equilateral triangle and shrink the entire plane towards that point, so that distances all become halved. That's one copy: the other two come from the other two corners. The rules for the Chaos Game correspond to these three transformations. The choice 'heads', 'tails', or 'edge' is like picking a transformation out of the bag. And the process of drawing dots, and moving them half-way towards the randomly chosen corner of the triangle, is exactly the one outlined above. So in fact it is no surprise to find that the Chaos Game draws a Sierpiński gasket.

That's what it was designed to do.

PASCALS EVERYWHERE

Fractals turn up in the most unexpected places. For example, they can be found in Pascal's triangle, named after the seventeenth-century mathematician-cum-philosopher Blaise Pascal. He doesn't really deserve all the credit: the idea was already known a lot earlier. But he does deserve *some* credit, because he developed some important applications of his triangle, among them some basics of probability theory. Pascal's triangle looks like this:

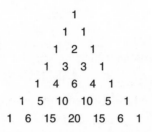

and so on. It arises in connection with the 'binomial theorem' in algebra: for example, if you work out

$$(x+y)^3 = x^3 + 3x^2y + 3xy^2 + y^3,$$

then the coefficients 1, 3, 3, 1 form the third row of Pascal's triangle. But that's by the by: what concerns us here is a beautiful pattern in Pascal's triangle, which lets us extend it as far as we wish. Each number in it (apart from the border of 1's) is equal to the sum of the two numbers one row above it, to the left and right. For example, the first 15 in row six is flanked, one row higher, by 5 and 10, like this:

$$5 \quad 10$$
$$15$$

and $5 + 10 = 15$. Using this rule, we can create Pascal's triangles with a thousand rows, or a million – given enough time and a fast computer.

Now, some of the numbers in Pascal's triangle are even, and some of them are odd. Obviously. But how can you tell which? The answer is elegant, and extremely surprising. To give you a hint of how strange it all is, here's a simpler question. If you drew a really big Pascal's triangle, say the first thousand rows, what proportion of the numbers in it would be even? About half? After all, up to any given limit half the numbers are even and the other half odd.

No. A lot more than half the numbers in the triangle are even. In fact, the larger the triangle, the closer the proportion comes to 100%. *Almost all* numbers in Pascal's triangle are even. Looking at a small triangle, this doesn't seem very likely, but I'll convince you that it must be true.

It helps to recast the problem in geometric terms. Draw Pascal's triangle as a grid of squares, like bricks in a triangular wall. Colour a square black if the corresponding number is odd, and white if it is even. We don't need to work out the exact numbers in Pascal's triangle in order to find out the colours. All we need is the symbolic rule that each number in the triangle is the sum of the two above it to the left and right:

$$a \qquad b$$
$$a+b$$

together with the information that

$$1 \text{ is odd,}$$
$$\text{odd} + \text{odd} = \text{even} + \text{even} = \text{even,}$$
$$\text{odd} + \text{even} = \text{even} + \text{odd} + \text{odd.}$$

Then the rule can be thought of as telling us how to get the colour of one brick from those of the two above:

$$\text{white} \qquad \text{white}$$
$$\text{white}$$

and

$$\text{white} \qquad \text{black}$$
$$\text{black}$$

and

$$\text{black} \qquad \text{white}$$
$$\text{black}$$

and

$$\text{black} \qquad \text{black}$$
$$\text{white.}$$

So all we have to do is colour all the squares along the two sides of the triangle black, and then colour squares white if the two above them are the same colour, black if not. It doesn't take long to fill in quite a big Pascal's triangle.

There's another, more sophisticated, way to describe all this. A number is even if it is congruent to zero (mod 2), and odd if it is congruent to 1 (mod 2). Because all the usual rules of algebra apply in modular arithmetic, you can build the entire triangle just using arithmetic (mod 2), following exactly the same addition rule.

Be that as it may, what do we actually get? We get a dramatic and intricate pattern of black and white upside-down triangles (Figure 86), which looks just like the Sierpiński gasket. This is no coincidence: it can be proved that the Sierpiński gasket pattern persists no matter how big a Pascal's triangle you draw.

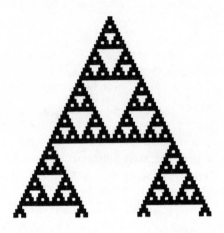

Figure 86 Pattern of odd and even numbers in Pascal's triangle.

Odd and even numbers occur equally often (give or take the odd extra one) in any selected range of whole numbers. You'd be forgiven for assuming that the same is true of the numbers in Pascal's triangle: half even, half odd. However, the probability of getting an even number in Pascal's triangle is the proportion that is coloured white in Figure 75, and the probability of getting an odd number is the proportion coloured black. For larger and larger numbers of rows in Pascal's triangle, these two probabilities are approximated better and better by the corresponding proportions of a Sierpiński gasket.

So what proportion of a Sierpiński gasket is white?

Consider how the gasket is constructed. Start with a black triangle of total area 1. Paint an upside-down triangle, one quarter the size, white. This leaves three smaller black triangles, each of area 1/4, and the remaining black area has shrunk from 1 to 3/4. Now paint an upside-down white triangle on each of these: the black area shrinks to $3/4 \times 3/4$. Repeat indefinitely. More and more of the gasket gets painted white, and the black area becomes $3/4 \times 3/4 \times \ldots \times 3/4$, which tends to *zero* as the number of stages becomes very large.

In other words, the black part of a Sierpiński gasket has total area 0, the white part has area 1. This means that *almost all numbers in Pascal's triangle are even*. So we've learned something surprising about Pascal's triangle, by thinking fractally about the Sierpiński gasket.

CALCULATOR CHAOS

We saw that Barnsley's Chaos Game also generates a Sierpiński gasket, but for a rather different reason. It's a pity he used the name 'chaos', though, because what his game uses is randomness. Chaos *looks* random – but it's not generated by a random process.

It is, however, closely related to fractals. Chaos is highly irregular dynamic behaviour generated by non-random rules. Fractals are the geometry of chaos – its visual manifestation.

Chaos arises in 'dynamical systems' – systems in which things change over time according to a fixed mathematical rule. The rule is of the following general type: 'If the state of the system now is *this*, then its state one instant into the future can be determined from *this* by calculating *that*.' At first sight, any such system is totally predictable. Given its state right now, we apply the rule to find what it will do one instant into the future. Then we apply the rule a second time to see what it will do one instant after that – two instants into the future. Then we apply the rule a third time to see what it will do one instant after that – three instants into the future.

This is what led Laplace to his rather ambitious vision of universal determinacy. However, dynamical systems are not as predictable as they first appear, and this is where chaos comes to the fore.

Chaos isn't anything exotic. It happens all the time – the only novelty is that nowadays we *notice* it. You can observe chaos on a pocket calculator.

Many calculators have an x^2 (square) button. (If not, $\times$ followed by $=$ has the same effect.) Pick a number between 0 and 1, such as 0.321, and hit the x^2 button. Do it again, over and over, and watch the numbers. What happens?

They shrink. By the ninth time I hit the button on my calculator, I get zero, and since $0^2 = 0$ it's no surprise that after that nothing very interesting happens.

This procedure is known as iteration: doing the same thing over and over again. A dynamical system works by iteration, too, with the same rule applied again and again over very tiny intervals of time. So the x^2

button is a good example of a dynamical system.

Try iterating some other buttons on your calculator. My description here assumes you always start with 0.321, but you can use other starting values if you want. Avoid 0, though. If you press the **cos** (cosine) button about fifty times, you'll find the mysterious number

$$0.739085133,$$

which thereafter just sits there. Again, the iteration of **cos** just settles down to a single value: it converges to a steady state.

The $1/x$ button (reciprocal) does something more interesting: the number switches alternately from

$$0.321$$

to

$$3.11526$$

and back again. The iteration is periodic, with period 2. That is, if you hit the button twice, you get back where you started.

The **exp** (exponential) button rapidly blows up to numbers so large that they exceed the calculator's capacity: you get the result 'E' – for 'error'.

Push all the buttons you've got: you'll find that the above outcomes – settle to a steady state, become periodic, or blow up – seem to be the only possible types of behaviour. Considering how many buttons the calculator has, this suggests a certain lack of variety in dynamical systems. But maybe this paucity of behaviour occurs because the buttons on a calculator are designed to do nice things? Nature may not be so accommodating. Let's invent new buttons. What about an $x^2 - 1$ button? To simulate it, hit the x^2 button and then $-1 =$. Keep doing this. You soon find you're cycling between 0 and -1, over and over again. The behaviour is just a periodic cycle again. There's nothing more predictable than a periodic universe: watch one cycle, and you know exactly what it will do for ever. Imagine the ease of weather forecasting if a given day of the week always had the same weather!

One last try: a $2x^2 - 1$ button. Now the numbers read

$$0.321, \quad -0.794, \quad 0.261, \quad -0.864, \quad 0.494, \quad -0.513, \quad -0.474, \ldots$$

They jump around a lot, without much pattern...

Aha!

This is chaos.

$2x^2 - 1$ is a simple enough rule. But the results of iterating it don't look so simple: in fact they look random.

Now try the $2x^2 - 1$ button again, but start with 0.322 instead of 0.321. It still looks random – and after just ten iterations it also looks completely different – so different that it's not obvious that the two lists are generated by the same calculator button. Even if you start with 0.3210001, the same thing happens – but now it takes about twenty iterations.

What makes this all the more remarkable is that while $2x^2 - 1$ is so weird, the superficially similar button $x^2 - 1$ is perfectly well behaved.

Why is this? The main reason is that the formula $2x^2 - 1$ magnifies tiny differences in the values of x, so that after a certain number of iterations values that started very close together move pretty much independently. Thus the behaviour is 'unpredictable'. The formula $x^2 - 1$ doesn't do this. It's as simple as that.

Chaos theorists call this phenomenon the butterfly effect. The name comes from a lecture given by the meteorologist Edward Lorenz, although he didn't actually use that phrase. He pointed out that weather is chaotic, and that it is therefore subject to the same sensitivity to tiny changes. If a butterfly flaps its wings today, said Lorenz, then a month later the world's weather will be quite different from what it would have been, had the butterfly *not* flapped its wings.

DON'T BLAME THE BUTTERFLY

Before you rush off to find the Chaos Butterfly and use it to hold the world to ransom, I should point out that in the real world we don't get to run the weather twice, nor do we have only one butterfly to worry about. Lorenz was dramatising the unpredictability of chaos. 'Prediction is very difficult – especially about the future.' So said the Nobel prizewinning physicist Niels Bohr. You don't need to be a Nobel prizewinner to know, from personal experience, that he was absolutely right.

Let's compare two everyday examples of dynamics: weather, and tides. You can find tables of the tides in diaries. But nobody includes a table of the weather. '6 June, full Moon, high tide at 7.42 a.m., sunny periods punctuated by light showers.' No, it doesn't ring true. Weather isn't that predictable.

Yet tides are.

Why are tides predictable, but weather not? Both tides and weather are governed by natural laws. The tides are caused by the gravitational attraction of the Sun and Moon, the weather by the motion of the atmosphere under the influence of heat from the Sun. The law of gravitation is not noticeably simpler than the laws of fluid dynamics, yet for weather the resulting behaviour seems to be far more complicated.

So, if it's not the complexity of the laws, why is there any difference at all? Because the laws for weather generate chaotic dynamics, and those for tides don't. A system can be unpredictable without being random. Laplace notwithstanding, it can be unpredictable even though it is deterministic.

Let's 'compare and contrast'.

A system is random if its future is independent of its past. It has no 'memory', and it is therefore *totally* unpredictable. Previous throws of an unbiased die provide no information about the next throw. Even if you've thrown twenty 6's in a row, the next throw is no more, and no less, likely to be a 6 than it was on every previous throw. There is always one chance in six of throwing a 6.

At the opposite extreme is the 'clockwork universe' behaviour envisaged by Isaac Newton in his laws of motion and gravity. In this view, the entire future – for ever – is completely determined by the present. The clockwork universe ticks for ever, and the only choice available to the deity is the positions of the cogwheels at the moment of creation. Behaviour like this is deterministic, and at first sight 'deterministic' seems to mean the same as 'predictable'.

Chaos sneaks in through a semantic distinction between what is predictable in high philosophical principle, and what you can actually do in practice. Yes, if you know the *exact* state of every particle in the universe now, you can in principle predict the future completely. But in practice you can't have exact information. Chaos occurs when any error, however tiny, grows as it propagates into the future, until eventually its effect becomes so large that the prediction is completely wrong. Chaotic systems are unpredictable in the long term. Just what this means depends on the system: a week for the weather, a microsecond for a turbulent fluid, a hundred million years for the solar system. But whatever the time-horizon may be, your prediction-telescope can't see beyond it.

ATTRACTORS

This description makes chaos seem useless: an obstacle to understanding, not an aid. Even if this were a valid interpretation, the universe wouldn't care very much: chaos is *there*, and it won't go away just because human beings dislike it. But there is a second, far more interesting, side to chaos – hidden patterns. Chaos is behaviour that looks random unless you know how to tease out the secret structure that reflects its deterministic origins. One of the most important secrets is that chaos is confined to

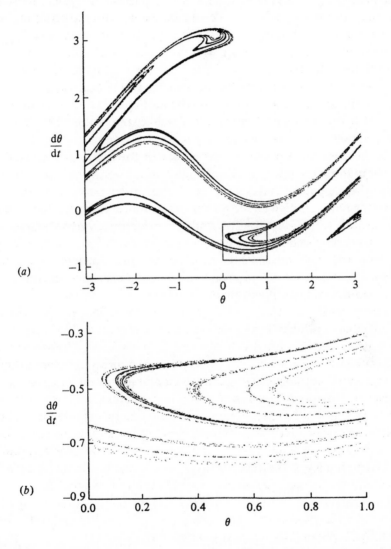

Figure 87 (a) The attractor of a forced damped pendulum. (b) Enlargement of the marked window to show the fractal fine structure. The horizontal coordinate is the angle at which the bob hangs, the vertical coordinate is its angular velocity.

well-defined ranges of behaviour, called *attractors*. The motion *on* the attractor is irregular and appears random, but the motions that confine the system *to* that attractor are predictable and regular.

As an analogy, imagine releasing a ping-pong ball on an ocean-covered planet. The ocean's surface is a chaotic place, with waves of all sizes buffeted by winds and pushed to and fro by currents. If you release the ball on the surface of the sea, it will follow a chaotic path. But if you release it from the bottom of the sea, it will float rapidly to the surface. Its upward motion will be relatively simple and predictable, though it will of course respond to some extent to the horizontal currents as well. Similarly, if you release it above the ocean, it will fall through the atmosphere until it reaches the surface. So the motion splits up into two parts. First, a highly regular motion *towards* the attractor (the ocean surface); then a much more irregular motion *on* it.

In this case the attractor is the surface of the entire ocean, which is rather big. It probably doesn't help much in understanding how the ball will move to know that it quickly reaches its attractor. But often the attractor is quite small. Figure 87 shows an attractor for a 'forced damped pendulum'. This is very like the familiar pendulum in a grandfather clock, but with some friction to slow it down and a rhythmic stimulus to keep it going despite that. Unlike the nice, periodic clock pendulum, it moves erratically. It is chaotic. But when you plot out a picture of the states that it takes up, you find a definite, non-fuzzy shape. That's the attractor. It's a very intricate shape, though – whereas the analogous shape for a grandfather clock pendulum is a single closed loop, corresponding to its simple periodicity.

The attractor of a system, with nice, regular dynamics, is very simple – a single point for steady states, a closed loop for periodic ones. A chaotic attractor is utterly different. It has structure on all scales.

It's a fractal.

CHAOS IN THE ASTEROID BELT

For a survey of the applications of chaos, try my book *Does God Play Dice?* To whet your appetite, I'm going to discuss an application to the long-term dynamics of our own dear solar system.

For centuries, astronomers have studied the regularities of the solar system – the orbits of the planets, the phases of the Moon, the paths of comets, the sunspot cycle. Now, equipped with the new mathematics of chaos, they can also study its irregularities. Some of these are important even for the most planetbound of human beings.

We don't normally worry much about the solar system: the Earth has been around for over four billion years, and we expect it to continue to provide a home for us for a good few years yet. So, perhaps, did the dinosaurs – until the notorious K/T meteorite crashed to Earth 65 million years ago and wiped them all out☞. And it's only a short time since Comet Shoemaker-Levy 9 made a spectacular series of impacts on the planet Jupiter, creating shockwaves bigger than the entire Earth. All of which makes our continued existence more fragile than we tend to imagine. The workings of chaos could at any time drop a large asteroid onto the Earth and wreak the kind of havoc described in Larry Niven and Jerry Pournelle's *Lucifer's Hammer* or Gregory Benford's *Shiva Descending*. Moreover, this cosmic destroyer need not be some un-suspected visitor from the depths of interstellar space. It could be any one of thousands of large rocks currently following regular and harmless orbits around the Sun at just the right distance between Mars and Jupiter. At any moment the precise, regular clockwork of the solar system could throw a very unexpected spanner into the machinery of our Earthly existence.

Paradoxically, it requires a very precise conjunction of circumstances for such a disaster to happen. It all depends on resonances – astronomical phenomena whose periods bear a simple numerical relationship to each other. For example, the Moon always faces the Earth, a $1:1$ resonance between its orbital and rotational periods. Mercury takes 87.97 days to revolve once round the Sun, and 58.65 days to rotate once on its axis. Two-thirds of 87.97 is very close to 58.65, so Mercury's orbital and rotational periods are in a $2:3$ resonance. Saturn has a large number of satellites, among them Hyperion and Titan. Hyperion takes 21.26 days to complete one orbit, and Titan takes 15.94. The ratio of the two periods is 1.3337, convincingly close to the ratio $4:3$.

Resonances are important because they imply that at regular intervals of time – the common period – the bodies in question bear *exactly* the same relationship to each other. This affects their dynamics. Some resonances are stable, and cause the bodies' motions to 'lock together' – just as the Moon's period of rotation on its axis is locked to its period of revolution around the Earth, so that we see only side of the Moon. Others are unstable, and cause wild behaviour. Which of these things happens depends on which resonance you've got.

It is resonances that could at any moment dump the equivalent of a gigatonne hydrogen bomb into our backyard. The effect is related to a long-standing astronomical conundrum, the gaps in the asteroid belt. The largest asteroid, Ceres, is about 940 km across. The smallest are little more than huge rocks, and there are hundreds of thousands of them. Most asteroids circle in the 'main belt' between the orbits of Mars and Jupiter, but a few come much closer to the Sun. The asteroid orbits are not spread uniformly between Mars and Jupiter. Their orbital distances from the Sun tend to cluster around some values and stay away from others. Daniel Kirkwood, an American astronomer who discovered this lack of uniformity in the 1860s, pointed out where the most prominent gaps occur. If a body were to encircle the Sun in one of these 'Kirkwood gaps', then its orbital period would resonate with that of Jupiter. Conclusion: resonance with Jupiter somehow perturbs any bodies in such orbits, and causes some kind of instability which sweeps them away to distances at which resonance no longer occurs. The special role of Jupiter is to be expected – it's so massive in comparison with the other planets.

Until recently, there were no mathematical methods for performing a long-term analysis of any of these resonances, but more powerful computers and new theoretical techniques now exist. The 3 : 1 resonance, for instance, is pretty well understood. The calculations show that an asteroid, orbiting at a distance that would undergo a 3 : 1 resonance with Jupiter, can follow a very irregular path. The eccentricity of its orbit – how fat or thin the orbit is – can change violently and almost at random. This is an astronomical example of chaos. The irregularities happen on a time-scale that's short by cosmic standards: about ten thousand years.

A main-belt asteroid whose orbit acquires eccentricity 0.3 or more becomes Mars-crossing, meaning that its orbit can cross that of Mars. Every time it does so, there's a chance that it will come sufficiently close to Mars to be hurled off into some totally different orbit. Until it was realised that chaos could generate high eccentricity, Mars-crossing was not a plausible mechanism for flinging asteroids at the Earth. Asteroids around the 3 : 1 Kirkwood gap were expected to stay well clear of Mars: there was no reason to expect a sudden change of eccentricity. But now there is such a reason, the mathematics of chaos. The 3 : 1 Kirkwood gap is there because Mars sweeps it clean, rather than being due to some action of Jupiter. Jupiter creates the resonance that causes the asteroid to become a Mars-crosser, then Mars kicks it away.

Jupiter creates the opening; Mars scores the goal.

Or maybe the touchdown.

Mars might *just* kick the asteroid in our direction (Figure 88). With

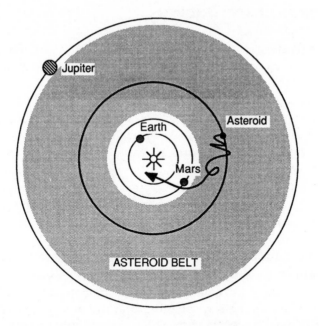

Figure 88   Cosmic soccer played with a chaotic asteroid. Jupiter centres from the corner – Mars scores?

Mars's help, the 3 : 1 resonance with Jupiter can transport rocks from the asteroid belt into Earth orbit, to burn up as meteorites in our planet's atmosphere.

And, if they're big enough, to burn us up instead.

Maybe it was just such a Martian 'invasion' that killed off the dinosaurs.

It's a sobering thought. Newton's beautiful, regular, clockwork laws are the rules for a random football game, played out by Mars and Jupiter on a cosmic battlefield. This game determines whether or not life continues to survive on Earth. Newton thought he'd ruled out divine intervention, but the unruly gods of the Enuma Elish are alive – and kicking. The chance that this celestial version of Russian roulette will prove fatal is admittedly very, very tiny – but it's there. The solar system is – literally – dicing with death.

Ours.

NOW FOR THE HAPPY ENDING ...

Since I made such a big fuss about chaos not being a Cassandra, I can hardly stop there, however dramatic an exit the total annihilation of humanity might offer.

Chaos has a good side, too.

For example – and this is just one of many – one consequence of the butterfly effect is that you can make big changes to a chaotic system with minimal effort. You can control it, easily. It sounds paradoxical, for chaotic dynamics looks completely out of control. But chaos has hidden patterns, and it is those patterns that make it controllable. An amazing new technique, known as chaotic control, has grown from this insight: the person most responsible is Jim Yorke, although there are several others.

One potential application is to build an intelligent heart pacemaker. It has been known for some years that certain kinds of heart disease involve the onset of chaotic dynamics. A tiny computer chip could be programmed to pick up these irregularities, and switch on a 'chaotic control' system that stabilises the heart back to more desirable regular rhythms. Current pacemakers do this with a big electrical current; an intelligent pacemaker would achieve better results with much less power, so its power supply would last far longer, avoiding the need to operate on the patient to put in a new 'battery'.

The same technique might be able to make passenger aircraft more efficient by getting rid of turbulence in the airflow over the wings. Thousands of tiny sensors would monitor the flow, and a computer system would exploit the chaotic behaviour of turbulence to manipulate thousands of tiny flaps, many times every second, cancelling the turbulence out again.

Even the space programme could benefit. Until recently, it was generally assumed that the most efficient orbit to send a package from the Earth to Moon – or from any world to any other – is a 'Hohmann ellipse'. This is an ellipse that surrounds both worlds, coming close to one of them at each end. However, it has now been discovered that a much more complicated orbit – one that is practically feasible only because of chaotic control methods – can do the same job using a lot less fuel. It takes longer – up to two years – so it wouldn't be used to send an astronaut.

But it could be used to supply a lunar colony.

Once more, the magic of the maze astounds us. When mathematicians first began to investigate chaos, they did so out of sheer curiosity. For years they were told that chaos was just a fad, a fashion, a load of media hype with no intellectual content and no uses whatsoever. Suddenly all of this turns out to be nonsense: the bug of fashion has bitten the ankles of its detractors, the potential applications of chaos seem inexhaustible.

When you explore the magical maze with an imaginative mind, marvels await you.

EXIT

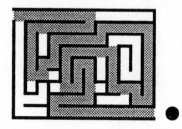

*B*efore you realise it, the fractal passageway of the yellow brick road opens *out into a meadow. A light drizzle is falling. You curse yourself for not bringing a raincoat. You console yourself with the thought that no one can predict the future in a chaotic universe – but you know that, even then, the sensible person can plan for contingencies. As long as weather remains on the same attractor, it will continue to look like weather.*

What weather is another matter altogether. But it certainly won't rain giant frogs.

A creature the size of a hippopotamus plummets to the ground a few metres away, bounces sluggishly, and settles, wobbling like a jelly.

It is a frog.

Maybe the universe is even less predictable than you thought. Clearly, you are not yet immune to the surreal influence of the magical maze.

You cast your mind back to your strange voyage, and the weird things you met along the way. And you realise that you barely scratched the surface of the magical maze. You wonder if one day you might go back, and explore it thoroughly. Depth First Search the entire magical maze!

A wild dream, a fantasy. The magical maze is not like the Minotaur's labyrinth. It cannot be explored in an afternoon, a month, or a millennium. Some parts are for ever inaccessible. But even those parts that are accessible are already of infinite extent. The explored regions are growing by the minute. In fact, in some curious manner, the entire magical maze is growing, all the time. Whatever the limits to mathematics, the maze is perpetually infused by new concepts from our even more magical universe.

You look for the white rabbit, but it has gone. So has the meadow, and the rain. You are at home, sitting in a chair, reading a book.

You close the book and

POINTERS

If you followed a ☛ in the main body of the magical maze, you should arrive here.

JUNCTION 1

1 'The Dormouse and the Doctor', in *When We Were Very Young* by A.A. Milne, Methuen, London, 1924, pp. 66–70. The opening verse goes:

> There once was a Dormouse who lived in a bed
> Of delphiniums (blue) and geraniums (red),
> And all the day long he'd a wonderful view
> Of geraniums (red) and delphiniums (blue).

PASSAGE 1 THE MAGIC OF NUMBERS

2 Martin Gardner, *The Annotated Snark*, Penguin, Harmondsworth, 1962, pp. 76–80.

3 The Butcher thereby makes a purely existential statement of a kind that has given mathematics a bad reputation in some quarters – not all of it undeserved. However, as we shall see in Passage 7, before solving a problem it may be important to know that an answer exists.

4 With hindsight, we should never have multiplied out $(x + 17) \times 992$ to get $992x + 16{,}864$. If we had left the expression alone, we'd have seen at once that $(x + 17) \times 992/992 = x + 17$. Practised algebraists, whose mathematical antennae feel their way slightly *ahead* of where the mathematician's thoughts are, would spot this simplification coming. But that's a minor tactical point.

5 The first great number theory book was Diophantus's *Arithmetica*, written around AD 250.

6 For chapter and verse on calendars of many cultures and historical

periods, and huge conversion tables, see Frank Parise, *The Book of Calendars*, Facts on File, New York, 1982. A more modern approach, based around computer programs, is the fascinating *Calendrical Calculations* by Nachum Dershowitz and Edward M. Reingold, Cambridge University Press, Cambridge, 1997. Their algorithms are available on the World Wide Wait (sorry, Web) at the website

http://emr.cs.uiuc.edu/home/reingold/calendar-book/index.html

For an accessible description of Fourier analysis see Morris Kline, *Mathematical Thought from Ancient to Modern Times*, Oxford University Press, New York, 1972, ch. 40. The crisis is discussed in E.T. Bell, *The Development of Mathematics*, McGraw-Hill, New York, 1945, ch. 13.

7   The adoption of the Gregorian calendar was sporadic. It was rapid in Catholic countries or provinces, slower in Protestant ones. When Sweden removed its ten days, it decided to do so gradually, and began by making 1700 a common year instead of a leap year. Unfortunately this process did not continue as planned, so it ended up one day wrong compared with the Julian calendar and nine days wrong compared with the Gregorian. It therefore reverted to the Julian calendar by giving 1712 – already a leap year – an extra day, 30 February, probably the first such date to occur since Julian times. The Gregorian calendar was not fully adopted in Sweden until 1844. In England it was adopted in 1750 and put into practice in 1752, by which time the equinoxes had slipped by a further day. Eleven days were therefore omitted (3 September was replaced by 14 September). The legendary cries of 'give us back our eleven days' seem to have been a rational response to various forms of financial hardship occasioned by the legal consequences of this loss, rather than by an irrational belief that people's lives had somehow been shortened. A consequence of this move that still remains to haunt us is the date of 6 April for the start of the financial year. It was originally 25 March, an inaccurate approximation to the spring equinox but one consistent with three-monthly accounting periods, including the all-important 25 December. With the addition of eleven days it became 5 April. In 1800 it became 6 April because a Julian leap year was omitted, but this anachronistic modification was not applied thereafter.

8   Nachum Dershowitz and Edward M. Reingold, *Calendrical Calculations,* Cambridge University Press, Cambridge, 1997. Here's an example of their algorithms, a relatively simple one. Given the

Gregorian month M, day D, and year Y, it calculates the number of days that have elapsed since 1 January in year 1 of the Gregorian calendar – counting that as day 1. (This date never occurred, historically, but it's a convenient point for conversions.)

The algorithm uses the 'floor function' $\lfloor x \rfloor$ = the greatest integer less than or equal to x, and goes like this. Compute:

1. $365(Y-1)$
2. $\lfloor (Y-1)/4 \rfloor - \lfloor (Y-1)/100 \rfloor + \lfloor (Y-1)/400 \rfloor$
3. $\lfloor (367M - 362)/12 \rfloor$
4. 0 if $M \leq 2$, –1 if $M > 2$ and Y is a leap year, and –2 otherwise
5. D

Now add them up, and you've got the number of days that have elapsed since 1 January in year 1 of the Gregorian calendar. The calculation has the following interpretation: (1) is the number of non-leap days in prior years; (2) is the number of leap days in prior years; (3) is a cunning formula for the number of days in prior months of year Y, based on the false but useful assumption that February has 30 days – as it probably did before Augustus got at it; (4) is a correction term to get February's length right; (5) adds in the number of days in the current month.

In the same spirit, here's an algorithm invented by Martin Stern, for finding the Gregorian date of the Islamic new year. It uses the mathematical operations $\lfloor \ \rfloor$, dec, mod and div. We already know about mod, div, and the floor function $\lfloor \ \rfloor$. For a number x, written as a decimal, the *decimal part* is dec $(x) = x - \lfloor x \rfloor$ everything after the decimal point. For example dec $(3.14159) = 0.14159$.

Let the year be Y by Muslim reckoning. Let

$$T = 0.970224Y + 621.5774.$$

Then the Gregorian year corresponding to Y is

$$G = \lfloor T \rfloor,$$

and the date of the Muslim new year is given by

$$D = \lfloor 365 \, \mathrm{dec}(T) \rfloor$$

in the sense that it occurs this number of days into the Gregorian year, so that $D = 1$ is 1 January, $D = 2$ is 2 January, ..., $D = 32$ is 1 February, and so on. Unfortunately there may be a glitch of up to one day, because of the interaction between Gregorian leap days and

Muslim intercalary days. To correct this, compute the day of the week w on which the Muslim new year falls by finding

$$r = Y \bmod 30$$
$$n = (11Y + 3) \operatorname{div} 30,$$
$$w = (4r + n + 2) \bmod 7.$$

Then $w = 0$ for Saturday, 1 for Sunday, 2 for Monday, and so on.

For example, let $Y = 1417$. Then $T = 1996.3848$, so that $G = 1996$ and $D = \lfloor 365 \times 0.3848 \rfloor = \lfloor 140.452 \rfloor = 140$. Bearing in mind that 1996 is a Gregorian leap year, this works out as 19 May. *But* maybe this is a day wrong. To find out, observe that 19 May 1996 is a Sunday. Then compute $r = 1417 \bmod 30 = 7$, $n = (11 \times 1417 \times 3) \operatorname{div} 30 = 15{,}590 \operatorname{div} 30 = 519$, $w = (4 \times 7 + 519) \bmod 7 = 547 \bmod 7 = 1$. The number 1 does indeed represent Sunday, so there is no error and the answer really is 19 May.

9 The nickname Fibonacci probably dates from the 1800s, and was invented by the French mathematician and populariser Édouard Lucas.

10 For this formula for Fibonacci numbers, and much else about them, see Ronald L. Graham, Donald E. Knuth, and Oren Patashnik, *Concrete Mathematics*, Addison-Wesley, Reading, MA, 1994, pp. 290–301.

11 We can measure 'how irrational' ϕ is by finding the best approximations to it by rational numbers p/q, and seeing how quickly the error – the difference between such a fraction and ϕ – shrinks towards zero as q increases. Number theorists have proved that for ϕ they shrink more slowly than they do for any other irrational number. See A.Ya Khinchin, *Continued Fractions*, Phoenix, University of Chicago Press, Chicago, 1964, p. 36.

12 Stéphane Douady and Yves Couder, 'La Physique des spirales végétales', *La Recherche* **24** (Jan. 1993), 26–35; 'Phyllotaxis as a self-organized growth process', in *Growth Patterns in Physical Sciences and Biology* (ed. J.M. Garcia-Ruiz *et al.*), Plenum, New York, 1993, pp. 341–352; 'Phyllotaxis as a self-organized growth process', *Physical Review Letters* **68** (1992) pp.2098–2101.

PASSAGE 2 PANTHERS DON'T LIKE PORRIDGE

1 T.H. O'Beirne, *Puzzles and Paradoxes*, Oxford University Press, Oxford, 1965, p. 2.

2   Answer to the pouring problem: Denote positions by the contents of the 8-, 5-, and 3-litre jugs arranged in that order. For example, 512 means 5 litres in the 8-litre jug, 1 in the 5-litre jug, and 2 in the 3-litre jug. One example of how the algorithm works (there are others because of the random choices involved) is given in Table 13.

Table 13

step	current position	possible moves	move made	comments
1	800	350	350	Starting position: make random choice
		503		Alternative move
2	350	800		Old position
		053	053	Random choice
		323		Alternative move
3	053	350		Old position
		503	503	Forced by algorithm
4	503	053		Old position
		530	530	Forced by algorithm
5	530	503		Old position
		233	233	Forced by algorithm
6	233	530		Old position
		503		Old position } backtrack
		053		Old position
7–10				Backtrack to position 2
11	350	As in 2	323	Forced by algorithm
12	323	350		Old position
		053		Old position
		503		Old position
		620	620	Forced by algorithm
13	620	323		Old position
		350		Old position
		602		Forced by algorithm
14	602	620		Old position
		503		Old position
		152	152	Forced by algorithm
15	152	602		Old position
		350		Old position
		143	143	Forced by algorithm
16	143	152		Old position
		503		Old position

053   Old position
440   440   Forced by algorithm
17   440   Finish

3   The Minotaur's argument is fallacious. For the construction of the universal sequence to work, the infinite 'list' of all mazes must be ordered in such a way that every element in it has finitely many pre-decessors. You can't always list an infinite set in this manner if you insist that a selected infinite subset appears first. For example, if you try to list the positive integers putting all odd integers first, then you get 1, 3, 5, 7, ..., and there's nowhere to put the even numbers. If you tag them 'on the end' to get

1, 3, 5, 7, ... (infinitely many odd numbers) ...; 2, 4, 6 ...,

then 2, for example, has infinitely many predecessors, so the above is not a list in the required sense.

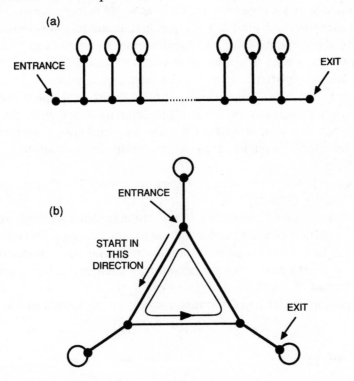

Figure 89   (a) Infinitely many mazes that are solved by LLLLL ...; (b) a maze that isn't.

In the same way, if you put all mazes that are solved by LLLLL ... first, which is an infinite list of mazes (Figure 89(a)), then you will miss out all mazes for which this sequence never gives a solution, such as Figure 89(b). The same holds for Theseus's variant RRRRR ..., of course.

PASSAGE 3 MARILYN AND THE GOATS

1   I can't believe that anybody doesn't know the details, but for all I know you're telepathing this from a mind-meld datacube in the year 3001, so here's a brief summary. Simpson was accused of murder. DNA evidence was prominent among that advanced by the prosecution. Lawyers and witnesses for both prosecution and defence argued about the interpretation, admissibility, and scientific status of that evidence. Ultimately the jury decided to acquit Simpson. Later he was successfully sued in a private prosecution by relatives of the deceased, and required to pay compensation. The real world is more complex than any mathematical model. My point is that the jury was obliged to think about the meaning of the probabilities presented to them, by 'expert witnesses', in conjunction with the use of DNA testing. I'm not discussing what conclusions they came to, or why.

2   Actually, that's no longer true: in 1997 a British trial opened up the question of *fingerprint* evidence again. Ordinary fingerprints, the kind that decorate your fingers with those tiny curlicues. I suspect the lawyers took inspiration from the problems surrounding DNA evidence.

3   Robert A.J. Matthews, 'The Interrogator's Fallacy', *Mathematics Today* **31** (Jan/Feb 1995), p. 3.

4   By 'context' here I mean what in the jargon is called a 'sample space'. It's a question of what model you use for the problem, and different models can give different probabilities. In general, Bayesian reasoning involves setting up a notion of 'prior probability', which is again a contextual choice.

5   Derivation of Matthews' formula: Let $P(A) = p$. By Bayes's theorem,

$$P(A \text{ given } C) = P(A \text{ and } C)/P(C),$$

and similarly

$$P(C \text{ given } A) = P(C \text{ and } A)/P(A)$$

But $(C \text{ and } A) = (A \text{ and } C)$, so we can combine the two equations to get

$$P(A \text{ given } C) = P(C \text{ and } A)/P(C)/P(A)$$

Moreover,

$$P(C) = P(C \text{ given } A)P(A) + P(C \text{ given not } -A)P(\text{not } -A),$$

since either A or not $-A$ must happen, but not both. Finally, P(not $-A$) $= 1 - P(A)$, so if we put $P(A) = p$, then P(not $-A$) $= 1 - p$. Putting all this together, we get the complicated-looking formula

$$P(A \text{ given } C) = P(A)/[P(A) + P(C \text{ given not } -A)/P(C \text{ given } A)P(\text{not } -A)].$$

If we replace $P(A)$ by p and $P(C$ given $A)/P(C$ given not $-A)$ by r, we get

$$P(A \text{ given } C) = p/[(p + r)(1 - p)],$$

as claimed.

6 Eugene P. Northrop, *Riddles in Mathematics*, Penguin, Harmondsworth, 1960.

7 Martin Gardner, *More Mathematical Puzzles and Diversions*, Bell, London, 1963, p. 165.

PASSAGE 4 THE SLIME MOULD SAGA

1 Ian Stewart, *Life's Other Secret*, Wiley, New York, 1998, ch. 4.

2 Jack Cohen and Ian Stewart, *The Collapse of Chaos*, Penguin, Harmondsworth, 1994; Ian Stewart, *Life's Other Secret*, Wiley, New York, 1998.

3 The Babylonian method is based on the formula $(x + y)(x - y) = x^2 - y^2$. To find ab, given two numbers a and b, let $x = (a + b)/2$ and $y = (a - b)/2$. Then $x + y = a$ and $x - y = b$. Therefore $ab = x^2 - y^2 = [(a + b)/2]^2 - [(a - b)/2]^2$.

4 Ian Stewart and Martin Golubitsky, *Fearful Symmetry*, Penguin, Harmondsworth, 1992, p. 43.

5 G. Polyá, 'Über die Analogie der Kristallsymmetrie in der Ebene' ['On the analogue of crystallographic symmetry in the plane'], *Zeitschrift für Kristallographie* 60 (1924), pp. 278–282.

6 Martin Golubitsky and Ian Melbourne, 'A symmetry classification of columns', preprint, Mathematics Department, University of Houston, 1996.

7 Syed Jan Abas and Amer Shaker Salman, *Symmetries of Islamic Geometrical Patterns*, World Scientific, Singapore, 1995, pp. 79–108.

8 Keith Critchlow, *Islamic Patterns*, Shocken, New York, 1976, p. 188.

9 Ian Stewart, *Life's Other Secret*, Wiley, New York, 1998, ch. 3.

10 Tibor Tarnai, 'Symmetry of golfballs', in *Katachi & Symmetry* (ed. T. Ogawa, K. Miuara, and D. Nagy), Springer-Verlag, Tokyo, 1996, pp. 207–214.

11 A simplification of Zhabotinskii's recipe, found by Jack Cohen and Arthur T. Winfree, works reliably even in front of students. It uses only four cheap components, having a shelf life of years – especially if they are refrigerated. It produces bromine, but not in dangerous amounts in an airy room. *Caution*: the mixture is moderately poisonous.

Make up four components:

A. 25 g sodium bromate, 335 ml water to dissolve, then 10 ml concentrated sulphuric acid.

B   10 g sodium bromide, water to 100 ml.

C   10 g malonic acid, water to 100 ml.

D   1,10-phenanthroline ferrous complex (available from Fisons, Lough-borough, UK).

Put 6 ml of solution A in a glass beaker, then add 0.5 ml of B, then quickly mix in 1 ml of C. Leave the brown mixture by an open window to lose bromine, until it is a pale straw colour or colourless (2–3 minutes if agitated in a flat dish). Add 1 ml of the redox indicator D, mix thoroughly, and pour into a 9 cm glass or plastic Petri dish on a white (preferably illuminated) background.

It will turn a patchy blue, then clear to a brown-red. Foci of blue will appear (you may have to wait up to 5 minutes) and grow into a series of concentric rings, expanding slowly. If the dish is shaken to restore homogeneity, the patterns reappear. Otherwise, do not jar or vibrate the dish. The effect lasts for 20–25 minutes. The experiment works very well on an overhead projector, for visibility in a classroom, but the rings may be fuzzy if the cooling fan is unbalanced.

12 Ian Stewart, *Life's Other Secret*, Wiley, New York, 1998, ch. 7.

13 Hans Meinhardt, *The Algorithmic Beauty of Sea Shells*, Springer-Verlag, Berlin, 1995.

14 Shigeru Kondo and Rihito Asai, 'A reaction–diffusion wave on the skin of the marine angelfish *Pomacanthus*', *Nature* **376** (1995), pp.765–768; Hans Meinhardt, 'Dynamics of stripe formation', *Nature* **376** (1995), pp.722–723.

PASSAGE 5 THE PATTERN OF TINY FEET

1 Whitman Richards (ed.), *Natural Computation*, MIT Press, Cambridge, MA, 1990.
2 Heat is really kinetic energy of component molecules: the body is said to get hotter when its component molecules jiggle around faster. But we don't usually build molecular structure into our Newtonian mechanical modelling, so it's easiest to think of heat as a third form of energy.
3 Oscillators can exist even in systems that do not conserve energy. In fact, many oscillators dissipate (lose) energy, which is then replaced from some external 'forcing' source. For many oscillators, the concept of energy is irrelevant.
4 For pictures of the gaits themselves see P.P. Gambaryan, *How Mammals Run*, Wiley, New York, 1974; Ian Stewart and Martin Golubitsky, *Fearful Symmetry*, Penguin, Harmondsworth, 1992, pp. 190–199.
5 J. Buck and E. Buck, 'Synchronous fireflies', *Scientific American* **234** (1976), pp.74–85.
6 The work of Kopell and Ermentrout is described briefly in Allyn Jackson, 'Lamprey lingo', *Notices of the American Mathematical Society*, **38** (1991), pp.1236–1239.

PASSAGE 6 TURING'S TRAIN SET

1 The twentieth century, of course, began on 1 January 1901, just as the third millennium begins, along with the twenty-first century, in 2001. But that's just being pedantic, isn't it? Trying to spoil all the millennial parties on 31 December 1999. Trouble is, it's also being *correct*. How annoying!
2 Fermat's Last Theorem, one of the most notorious open questions in the whole of mathematics, one that survived unscathed for more than 350 years, was finally proved by Andrew Wiles in 1994. Pierre de Fermat (1601–1665) was a French lawyer, but he pursued mathematics as a hobby and was so good at it that he ranks among the all-time greats. Nearly half a century before Newton, he worked out many of the basic ideas of calculus, but his most significant achievements were in number theory. It was known to the ancient Greeks that there are infinitely many different Pythagorean triples – whole numbers that

can form the sides of a right triangle. Pythagoras's theorem tells us that such numbers x, y, z must satisfy the equation $x^2 + y^2 = z^2$. Well-known examples are $3^2 + 4^2 = 5^2$ and $5^2 + 12^2 = 13^2$. Fermat wondered whether the same kind of thing could occur for cubes, fourth powers, and so on; and he convinced himself that it could not. In the margin of his copy of the *Arithmetica* by Diophantus, he wrote:

> To resolve a cube into the sum of two cubes, a fourth power into two fourth powers, or in general any power higher than the second into two of the same kind, is impossible; of which fact I have found a remarkable proof. The margin is too small to contain it.

Fermat was asserting that the 'Fermat equation' $x^n + y^n = z^n$ has no integer solutions when $n \geq 3$, other than trivial solutions where one number is zero. Fermat's 'remarkable proof' has never been found, and experts generally believe that whatever Fermat had in mind it must have contained an error. More than two hundred years elapsed before the first really big inroad on the Last Theorem was made, by Ernst Kummer. By devising an entirely new theory of 'ideal numbers' he was able to prove Fermat right for any $n \leq 100$, except perhaps for $n = 37$, 59, and 67. Later mathematicians polished off these possible exceptions and pushed the limits out to the 150,000th power.

Meanwhile, in 1922, the British mathematician Leo Mordell came up with another conjecture, one of whose consequences is that for any $n \geq 3$ there is at most a finite number of different solutions to the Fermat equation. Mordell's intuition was vindicated in 1983 by Gerd Faltings. Soon afterwards, D.R. Heath-Brown proved that Fermat's Last Theorem is true for 'almost all' n – if there are any exceptional values of n, for which solutions do exist, then those values must thin out enormously as n becomes large.

In the 1980s and early 1990s it became clear that if another, more powerful, more general, and more credible conjecture – called the Taniyama–Shimura conjecture – is true, then so is Fermat's. The Taniyama–Shimura conjecture centres on a rich area of number theory, 'elliptic curves'. In 1993 Wiles announced a proof of a special case of the Taniyama–Shimura conjecture, good enough to imply Fermat's Last Theorem, but soon an error was found. He managed to repair the proof in 1994.

For further details, see Ian Stewart, *From Here to Infinity*, Oxford University Press, Oxford, 1996; Simon Singh, *Fermat's Last Theorem*, Fourth Estate, London, 1997.

3   Douglas Hofstadter, *Gödel, Escher, Bach: An Eternal Golden Braid*, Penguin, Harmondsworth, 1980.

4   Modern generalisations of the notion of computation may force a rethink here soon.

5   Adam Chalcraft and Michael Greene, 'Train sets', *Eureka* 53 (1994), pp.5–12.

6   Elwyn R. Berlekamp, John H. Conway, and Richard K. Guy, *Winning Ways*, Vol. 2, Academic Press, New York, 1982, p. 817.

7   This marvellous image comes from Douglas Hofstadter, *Gödel, Escher, Bach: An Eternal Golden Braid*, Penguin, Harmondsworth, 1980.

PASSAGE 7 QUEEN DIDO'S HIDE

1   How do we know that these changes to the angle really do increase the area?

The way we change the curve is this: we leave the shaded regions as they were, apart from moving them around a bit, and we open up or squash the white triangle until the top angle is 90º. What I'm claiming is that this angle makes the triangle's area maximal. Notice that when we open up or squash the triangle, the two sloping sides don't change in length – all that changes is the base. So what I'm really claiming is this: *if you are given two sides of a triangle, then its area is greatest when the angle between them is a right angle.*

Figure 90 shows various possible positions for such a triangle, including one where the angle is a right angle, one where it is less,

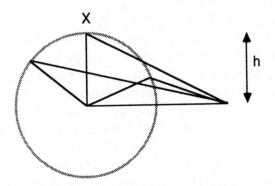

Figure 90   Maximising the area of a triangle when two sides are given.

and one where it is more. Now the area of such a triangle is half the base times the height (h). So we want to move the end marked X to the highest possible position. Since X wanders round a circle as we vary the angle, we want to determine the point of the circle that is highest. This is of course the point that lies along a line at right angles to the base of the triangle, and that's exactly what we want to establish.

2   The fact that the angle in a semicircle is 90º is so well known that I won't prove it. Unfortunately, what we need is the converse: if all such angles are 90º then the curve is a semicircle. That is, *if* the curve is a circle, then the part lying above the line AB is a semicircle, and then angle ACB must be 90º. Now this is very similar to what we already know, but the *if ... then* is the wrong way round. We know that 'if it's a circle, then angle ACB is 90º for any C'. Unfortunately, what we have to prove is that 'if angle ACB is 90º for any C, then it's a circle'.

That's not the same statement. Compare 'if it's raining, then my garden gets wet' with 'if my garden gets wet, then it's raining'. The first is true. The second could be false – for example, I may be watering the garden with a hose. So the two statements aren't logical equivalents of each other.

To fix things up, we compare our (half-)curve with a (semi)circle and show that there's no difference. Suppose, then, that our half-curve is not a semicircle. Then we can find a diameter AB and a point C such that C does not lie on the circle with diameter AB. This means that AC cuts the circle at a point D *different from* C. Now, we know that angle ACB is 90º because we've proved that our curve has that property. We *also* know that angle ADB is 90º, because that's how circles behave. So the line CB must be parallel to DB, since both cut the same line at right angles. However, the two lines CB and DB meet at B, whereas parallel lines don't meet at all.

There's nothing wrong with the logic, so our initial assumption has to be at fault. What was it? That the half-curve isn't a semicircle. Conclusion: actually, it *is* a semicircle.

3   John H. Halton, 'The shoelace problem', *The Mathematical Intelligencer*, **17**(4), 1995, pp.36–40.

4   The lengths are

> *American*: $g + 2n\sqrt{(d^2 + g^2)}$.
>
> *European*: $ng + 2\sqrt{(d^2 + g^2)} + (n-1)\sqrt{(4d^2 + g^2)}$.
>
> *Shoe-store*: $ng + n\sqrt{(d^2 + g^2)} + \sqrt{(n^2d^2 + g^2)}$.

5   Richard Courant and Herbert Robbins, *What is Mathematics?*, revised by Ian Stewart, Oxford University Press, New York, 1996, pp. 354, 392.

6   In 1996, Newton Da Costa and C. Doria announced a proof that the 'P $\neq$ NP' problem is undecidable. If so, it's impossible to prove that *any* NP-complete problem really is hard. It's also impossible to prove that any NP-complete problem is easy. Their proof has not appeared at the time of writing.

PASSAGE 8 GALLERY OF MONSTERS

1   For a more extensive discussion of nearly everything in this final passage of the magical maze, and a lot more, see Ian Stewart, *Does God Play Dice?* (2nd edn), Penguin, Harmondsworth, 1997.

2   If $x = e^y$, where $e = 2.718 \ldots$ is the base of natural logarithms, then $y = \log x$. The main property of logarithms that we use here is

$$\log(ab) = \log a + \log b,$$

together with its consequence

$$\log(a^c) = c \log a.$$

3   See Walter Alvarez and Frank Asaro, 'An extraterrestrial impact', *Scientific American* (Oct 1990) pp. 44–52. For a contrary view, see Vincent E. Courtillot, 'A volcanic eruption', *Scientific American* (Oct 1990), pp. 53–60.

DIRECTORY